# Farm Management

With Information on the Business, Marketing and Economics of Running a Farm

By

John W. Carncross
L. A. Bevan
W. R. Stone

Copyright © 2013 Read Books Ltd.
This book is copyright and may not be
reproduced or copied in any way without
the express permission of the publisher in writing

British Library Cataloguing-in-Publication Data
A catalogue record for this book is available from the
British Library

# Farming

Agriculture, also called farming or husbandry, is the cultivation of animals, plants, or fungi for fibre, biofuel, drugs and other products used to sustain and enhance human life. Agriculture was the key development in the rise of sedentary human civilization, whereby farming of domesticated species created food surpluses that nurtured the development of civilization. It is hence, of extraordinary importance for the development of society, as we know it today. The word *agriculture* is a late Middle English adaptation of Latin *agricultūra*, from *ager*, 'field', and *cultūra*, 'cultivation' or 'growing'. The history of agriculture dates back thousands of years, and its development has been driven and defined by vastly different climates, cultures, and technologies. However all farming generally relies on techniques to expand and maintain the lands that are suitable for raising domesticated species. For plants, this usually requires some form of irrigation, although there are methods of dryland farming. Livestock are raised in a combination of grassland-based and landless systems, in an industry that covers almost one-third of the world's ice- and water-free area.

Agricultural practices such as irrigation, crop rotation, fertilizers, pesticides and the domestication of livestock were developed long ago, but have made great progress in the past century. The history of agriculture has played a major role in human history, as agricultural progress has been a crucial factor in worldwide socio-economic change. Division of labour in agricultural societies made (now) commonplace specializations, rarely seen in hunter-gatherer cultures, which allowed the growth of towns and cities, and the complex societies we call civilizations. When farmers became capable of producing food beyond the needs of their own families, others in their society were freed to devote themselves to projects other than food acquisition. Historians and anthropologists have long argued that the development of agriculture made civilization possible.

In the developed world, industrial agriculture based on large-scale monoculture has become the dominant system of modern farming, although there is growing support for sustainable agriculture, including permaculture and organic agriculture. Until the Industrial Revolution, the vast majority of the human population laboured in agriculture. Pre-industrial agriculture was typically for self-sustenance, in which farmers raised most of their crops for their own consumption, instead of cash crops for trade. A remarkable shift in agricultural practices has occurred over the past two centuries however, in response to new technologies, and the development of world markets. This also has led to technological improvements in agricultural techniques, such as the Haber-Bosch method for synthesizing ammonium nitrate which made the traditional practice of recycling nutrients with crop rotation and animal manure less important.

Modern agronomy, plant breeding, agrochemicals such as pesticides and fertilizers, and technological improvements have sharply increased yields from cultivation, but at the same time have caused widespread ecological damage and negative human health effects. Selective breeding and modern practices in animal husbandry have similarly increased the output of meat, but have raised concerns about animal welfare and the health effects of the antibiotics, growth hormones, and other chemicals commonly used in industrial meat production. Genetically Modified Organisms are an increasing component of agriculture today, although they are banned in several countries. Another controversial issue is 'water management'; an increasingly global issue fostering debate. Significant degradation of land and water resources, including the depletion of aquifers, has been observed in recent decades, and the effects of global warming on agriculture and of agriculture on global warming are still not fully understood.

The agricultural world of today is at a cross roads. Over one third of the worlds workers are employed in agriculture, second only to the services sector, but its future is uncertain. A constantly growing world population is necessitating more and more land being utilised for growth of food stuffs, but also the burgeoning mechanised methods of food cultivation and harvesting means that many farming jobs are becoming redundant. Quite how the sector will respond to these challenges remains to be seen.

# FARM MANAGEMENT

## John W. Carncross

*Problems of farm management and farm prices have been the concern of John W. Carncross for nearly 18 years. As associate agricultural economist at the New Jersey Experiment Station, he spends much time visiting farmers, studying their practices, and analyzing their accounts and methods of management. Out of this experience have come many reports and publications which, based on producers' records, highlight fundamentals of efficient farm management. General farms, as well as specialized enterprises devoted to the production of vegetables, apples and peaches, milk and dairy products, eggs and poultry, and potatoes, are among the kinds of businesses studied and analyzed by Mr. Carncross. He has studied the economic influence of power machinery on various types of farms, and the benefits of soil conservation practices. These and other subjects have been covered in many technical reports and popular articles. Mr. Carncross is a native of New York, a graduate of Cornell University.*

The operator of a commercial farm is engaged in a highly competitive enterprise. His farm is a business unit or "factory" for production. If the farmer is successful, it is due in no small measure to his knowledge of economic principles involved in the organization and operation of the farm as an efficient producing unit.

As an entrepreneur, the farmer combines labor and capital in the hope of profits, but always with the risk of losses. In addition to his own labor, the farmer usually has the help of members of his own family and very often of hired labor.

How to make profits and avoid losses calls for keen judgment in the selection and combination of crops and livestock as a production unit, and the skillful application of farm experience and techniques to every step in the processes of production. Each of these involves economic considerations which have a direct bearing on the efficiency of production and, therefore, on the success of the farm business as a whole.

Such considerations as size of business, crop yields, rate of animal production, labor efficiency, and price levels are among the determinants of the success of the farm as a business venture. All are inter-related. Achieving a desirable adjustment among them is the essence of good farm management.

There can be no hard and fast rules, applicable to all farms in all circumstances, which blueprint the means for achieving efficient management. Not only do we have many types of farming but, within the many types, each farm presents its own peculiar problems of management. This is a fact to remember in appraising all that is set forth in the following discussion of fundamentals underlying good farm management. Such management reflects keen thinking and sound planning on the farm—thinking and planning based on the farm's resources, limitations, and available markets. If this discussion assists in the guidance of that thinking and planning, it will have served its purpose.

### Farm Labor Income Defined

Size of the farm, rates of crop and animal production, labor efficiency, and the other factors in good farm management discussed in this chapter—all have a direct bearing on labor income, a term which requires definition since it will be used frequently here.

Labor income is the cash the farmer has at the end of the season or year after deducting expenses (including a charge for interest on his investment) from total receipts. It represents income which can be considered as payment for the farmer's own labor and management. In addition, the farmer has the use of the farm dwelling (this value is comparable to the cost of renting or owning a home in the city), and he also has the value of such products—eggs, milk, vegetables, and the like—as are available from the farm for the family.

### Size of Business as Affecting Receipts and Profits

In spite of all that has been said and written about large-scale, corporation farming, agriculture in the United States consists largely of many thousands of relatively small units, with the operator and his family serving as the principal source of labor. Many of these farms, of course, employ some additional help during part or all of the year.

## FARM MANAGEMENT

That good management calls for a farm of a desirable size can be said with all assurance. Size of the farm vitally affects volume of production; volume of production vitally affects receipts and profits. But there can be no one answer to the question, "How large a farm shall I buy?"

Some light is thrown on the question when it is narrowed down to a particular type of farming in a specified area, and when the prospective purchaser provides information on such details as his financial resources, reasons for seeking a farm, training and experience, and standard of living acceptable to himself and family.

In general—and this definitely is a generalization—a commercial farm should be of sufficient size to require the full employment of two men or their equivalent.

The individual seeking the most light on the question would do well to talk with the County Agricultural Agent of the county in which he wishes to buy a farm. Such a talk will be more helpful before, rather than after, buying.

Some light on the relationship between the size of a farm business and labor income is apparent in results of a study (Table 1) made of records kept on a number of dairy farms in Hunterdon County, New Jersey. The relationship is one that applies to dairy farms generally.

TABLE 1

SIZE OF BUSINESS, OR NUMBER OF DAIRY COWS, RELATED TO LABOR INCOME, 1940

(Average per Farm)

| Farms with | 10 to 20 Cows | More than 20 Cows |
|---|---|---|
| Number of Cows | 14 | 30 |
| Labor Income | −$137 | $867 |
| Total Receipts | $3,966 | $8,476 |
| Total Capital | $14,352 | $23,577 |
| Total Acres in Farm | 106 | 166 |
| Acres in Crop Land | 76 | 107 |
| Number of Farms | 28 | 15 |

The 28 farms keeping from 10 to 20 milk cows each (an average of 14) reported average gross sales of $3,966. These farms reported an average labor income of minus $137; i.e., total receipts were $137 less than total costs of operation.

## FUNDAMENTALS OF FARMING

The 15 farms keeping more than 20 cows each (an average of 30) reported average gross sales of $8,476 and labor income of $867.

The smaller farms are not of a sufficient size to justify the full time employment of two men, although many of the jobs involved in crop production can be most efficiently performed when two

*A large volume of business is apparent in this farm.*

or more men are employed. The 30-cow dairies, on the other hand, would, on the average, use the labor of two men the year round, and additional help during the crop-harvesting period.

The limitations of small size are further illustrated (Table 2) by figures from poultry farms, where the raising of chickens and the sale of eggs were the principal sources of income.

The 10 farms keeping less than 500 birds and averaging 339 closed the year with a labor income of minus $864. Because the

TABLE 2

SIZE OF BUSINESS, OR NUMBER OF HENS, RELATED TO LABOR INCOME ON POULTRY FARMS, 1938

| Size of Flock | Under 500 Birds | 500– 1000 Birds | 1000 Birds and Over |
|---|---|---|---|
| Average Number of Birds[1] | 339 | 783 | 1232 |
| Labor Income | –$864 | –$368 | $566 |
| Total Farm Receipts | $1,274 | $3,054 | $5,202 |
| Total Capital | $7,858 | $10,330 | $11,865 |
| Total Acres in Farm | 67 | 53 | 45 |
| Acres of Crop Land | 57 | 42 | 28 |
| Number of Farms | 10 | 14 | 16 |

[1]Number of Hens and Pullets in a laying flock at beginning of year.

## FARM MANAGEMENT

size of their flocks was small, these poultry farms could not produce sufficient income to cover costs of production, and there was no return to the operator for his labor. From 300 to 500 birds will not occupy the full time of one man's labor, and certain of the fixed overhead expenses which would be very little more for a considerably larger flock, run up a large fixed cost per bird for the small flock.

The larger flocks, averaging 1232 birds for the 16 farms, produced labor incomes of $566 on the average. Investment in the larger farms was at the rate of about $10 per bird as compared to nearly $25 per bird for the smaller-size farms.

Large size of business alone is no guarantee of a favorable labor income. It does, however, enhance the possibilities.

Misner and Lee of Cornell University Agricultural Experiment Station showed the same principle of relation of the number of layers to labor income for a 5-year period (1929–1933) in New York State. Small flocks averaging 524 layers produced a labor income of $435, medium flocks of 894 layers netted $851, and large flocks of 1979 layers averaged a labor income of $1,816.

Total receipts reflect the size or volume of the farm business. Most types of farming have fixed costs which require a large volume of business turnover for efficient operation. This general relation of total farm receipts to labor income is indicated (Table 3) in the records of a number of fruit and vegetable farms.

TABLE 3

RELATION OF TOTAL RECEIPTS PER FARM AND OTHER FACTORS TO LABOR INCOME ON FRUIT AND VEGETABLE FARMS

Average per Farm per Year

| Receipts per Farm | Lowest Third | Medium Third | Highest Third |
|---|---|---|---|
| Total Farm Receipts | $2,161 | $7,696 | $15,641 |
| Labor Income | -$1,218 | $158 | $3,153 |
| Crop Acres | 64 | 98 | 124 |
| Acres of Apples | 10 | 25 | 53 |
| " " Peaches | 9 | 14 | 17 |
| " " Vegetables | 18 | 42 | 38 |
| Yield of Apples per Acre-Bu. | 67 | 117 | 168 |
| Total Capital Invested | $11,408 | $21,137 | $29,052 |

Study made in Burlington County, New Jersey, 1934.

## FUNDAMENTALS OF FARMING

Farms reporting the highest total receipts, averaging $15,641 per farm, made labor incomes of $3,153 compared to an average labor income of minus $1,218 for the third of the farms with lowest farm receipts, averaging only $2,161. A further study of table 3 reveals that not only was there a difference in actual size of farms as measured in terms of crop acres, but there was some difference in the diversity of business, the larger farms having a greater proportion of the acres in apples.

Another outstanding difference associated with total receipts was the yield of apples per acre, which was considerably higher for the larger-size farms. Obtaining good crop yields is a very important part of good farm management.

### RATES OF CROP AND ANIMAL PRODUCTION

Rate of crop and animal production is one of the most important factors in efficiency of farm operation, through its effect in reducing the cost of a unit of output. The part which good crop yields play in increasing the earnings of the operator for some central New Jersey specialized potato farms is shown in Figure 1 and Table 4.

The 90 farms were divided into three groups with highest, intermediate, and lowest yields. The 30 farms with the highest yields had an average labor income of $1,525 and their rate of potato production per acre was 33 per cent above the average of the 90 farms. The intermediate group of farms with about average yields had a labor income of only $37.00, while the farms with lowest yields, averaging 24 per cent below the average of all the farms, had a labor income of minus $1,248. These farms averaged about the same in total acres and there was little difference between the groups in acreage of potatoes per farm. The farms with highest yields had total receipts from potatoes of $8,669, or nearly $3,400 per farm more than the group with lowest yields.

Higher net incomes are also obtained on vegetable farms where the rates of crop yields are above the average of the community. An average labor income of $374 was obtained on specialized vegetable farms in southern New Jersey, where the crop index or rate of production averaged about one-third higher than for all of the farms in the study as shown in table 5.

# FARM MANAGEMENT

## YIELD OF POTATOES PER ACRE AND LABOR INCOME ON SPECIALIZED POTATO FARMS[1]

### FIGURE 1

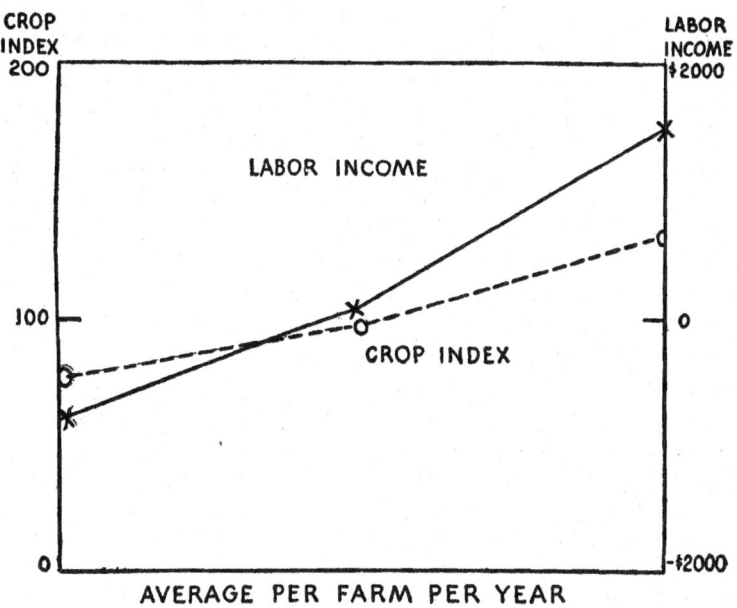

AVERAGE PER FARM PER YEAR

TABLE 4
Average per Farm per Year

| Potato Yield Index | Lowest 30 Farms | Intermediate 30 Farms | Highest 30 Farms |
|---|---|---|---|
| Average of Group (100=Aver. Yield of 90 Farms) | 76 | 99 | 133 |
| Labor Income | -$1,248 | $37 | $1,525 |
| Yield of Potatoes per Acre— 100-pound bags | 79 | 103 | 131 |
| Acres of Crop Land | 104 | 114 | 113 |
| Acres of Potatoes | 56 | 59 | 57 |
| Receipts from Potatoes | $5,295 | $7,234 | $8,669 |
| Total Farm Receipts | $6,663 | $8,852 | $10,456 |
| Total Capital | $19,588 | $19,236 | $22,913 |

[1] Farms with 60% or more of total receipts from potatoes. Monmouth County, New Jersey, 1939.

7

# FUNDAMENTALS OF FARMING

TABLE 5

RELATION OF CROP YIELD RATES OR INDEX
ON VEGETABLE FARMS AND LABOR INCOME

(Average yield of 37 farms=100)

Average per Farm per Year

| Crop Yield Rate | Below Average | Above Average |
|---|---|---|
| Crop Index | 77 | 132 |
| Labor Income | −$468 | $374 |
| Acres of Crop Land | 46 | 52 |
| Total Farm Receipts | $2,629 | $6,820 |
| Total Capital | 4,798 | 6,559 |
| Total Receipts per Acre of Crop Land | $62 | $118 |
| Fertilizer Expenses per Acre of Crop Land | $10 | $18 |

The group of farms with yields averaging about one-fourth less than for all of the farms had a minus $468 labor income. The total receipts per acre of crop land on these lower-yielding farms was $62, compared with $118 or nearly twice as much for the higher-yielding farms. The lower-yielding farms had an average expenditure of $10 per acre of crop land for fertilizer, as compared with $18 or nearly twice that amount for the higher-yielding farms. There are certain expenditures in connection with the cost of production which are incurred, up to harvest, and which are about the same per acre regardless of the yield obtained. These include such items as plowing and fitting the land, application of fertilizer, care of the crop, use of equipment, taxes and interest, etc. In other words, these are the more fixed items of cost of production. This general principle of the effect of higher yields in reducing unit costs of production is illustrated in the case of tomatoes for processing shown in Figure 2 and Table 6.

It will be noted that the group of farms with an average yield of 3.5 tons per acre of tomatoes for manufacture had a total cost of production per ton of nearly $25, compared with a cost of $11.67 per ton for the group having an average yield of 11 tons per acre. The group of farms with lowest yields (3.5 tons) showed a return per hour of man labor of minus $0.16. This means that there was not enough money obtained out of the crop to pay anything at all for man labor, nor was there enough to pay the other costs of production. The higher-yielding farms, on the other hand, paid for all of the expenses of production and pro-

## FARM MANAGEMENT

RELATION OF CROP YIELDS FOR TOMATOES FOR MANUFACTURE
AND COSTS PER TON, 1940[1]

FIGURE 2

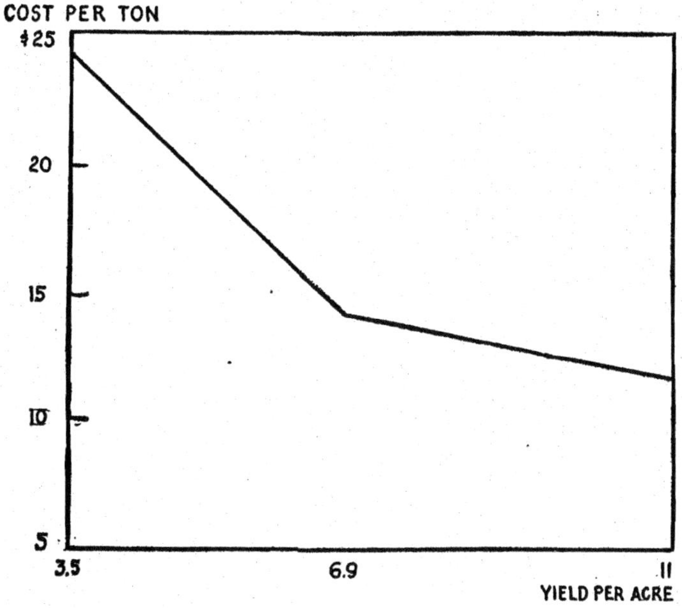

COST PER TON vs YIELD PER ACRE

TABLE 6

| Yield per Acre (Group Range) | Under 5 Tons | 5–9.99 Tons | 10 Tons and Over |
|---|---|---|---|
| Average Yield of Group | 3.5 | 6.9 | 11.0 |
| Cost of Production per Ton | $24.86 | $14.44 | $11.67 |
| Returns per Hour of Man Labor | −.16 | .33 | .48 |
| Fertilizer Applied—Lbs. per A. | 1088 | 1248 | 1392 |
| Number of Fields | 26 | 55 | 11 |

[1] 1940 Tomato Study in New Jersey.

vided a return of $0.48 per hour for every hour of man labor spent on the crop.

The type of soil and its adaptation to the particular crop and practices followed in producing the crop largely influence the yields which can be obtained. For example, the vegetable industry in southern New Jersey is located, for the most part, on sandy and loamy soils which are fairly well adapted to truck-crop production. There are many acres, however, on which it is not eco-

## FUNDAMENTALS OF FARMING

nomically practical to produce vegetables, because of the lightness of the soil, the low state of fertility, and the high expense which would be involved in adding and maintaining organic matter and nutrients necessary for obtaining a favorable yield. On many farms, however, the failure to follow good crop practices in the way of sufficient applications of fertilizer, lime, and manure results in low or uneconomical crop yields. It will be noted in the case of tomatoes for manufacture that the high-yielding group used about 300 pounds per acre more of fertilizer than the low-yielding group.

*Borrowing or lending money for the operation of a disease-infected, run-down apple orchard having only poor varieties is not good business.*

Farm production, under modern conditions, is very complicated, and there are many detailed operations on the farm which require knowledge and judgment on the part of the practical farm operator which are not so apparent to the layman. The following of a routine program is essential to the maximum production of animals, and the timing of planting and caring for crops is necessary to highest yields. In the study of tomatoes for processing, it was found that the fields which were planted before the 15th of May produced a yield of 7.8 tons, while the fields which were planted May 15th or later produced an average yield of 6 tons per acre. The cost of production per ton on the early-planted fields was $13.03, while on the late-planted fields tomatoes were produced at a cost of $16.95 per ton.

The economic feasibility of the production of individual crops can be determined in part from a study of the yields per acre necessary to meet expenses at anticipated prices. An illustration

## FARM MANAGEMENT

of the yields necessary to meet expenses at different farm prices is shown for onions in Table 7.

TABLE 7

AVERAGE YIELD REQUIRED TO MEET EXPENSES OR BREAK EVEN BASED ON 1939 ESTIMATED COSTS

*Onions*

| If Farm Price Received Is: | Yield per Acre |
|---|---|
| $0.60 per bag (50 ℔) | 344 bags |
| .65   "    "    " | 302  " |
| .70   "    "    " | 270  " |
| .75   "    "    " | 244  " |
| .80   "    "    " | 223  " |
| Cost to harvest per 50 ℔ bag | 23 cents |
| Cost up to harvest per acre | $127 |
| 1939 State-average price | $0.65 |
| 10-year State-average yield | 276 bags |

With a price of 60 cents per bag—the existing price in 1939—a yield of 344 bags would be required to break even. At a price of 80 cents per bag, an estimated yield of 223 bags would be required to meet expenses. If the average price which a grower could anticipate for his crop was 65 cents per bag, he would know that it would be necessary to obtain a yield of around 300 bags per acre to meet expenses. Further, if he knows that under his soil and production conditions he can average only around 225 bags per acre, he would do well to make a similar analysis of other crops to determine possibly more favorable enterprises, or look into the possible factors which may be limiting his yield and which could be economically remedied.

The economic advantage of higher rates of production can also be shown in the case of dairy herds. To illustrate this point, data are shown in Table 8 comparing the average of the five lowest and the five highest of 43 producing herds in 1940. The herds with lowest production, averaging 207 pounds of butterfat and 5479 pounds of milk, had a labor income of $386, while the higher-producing herds, with a production per cow of 363 pounds of butterfat and 9113 pounds of milk, had a labor income of $1293 per farm. It should be pointed out, however, that high rates of production will not themselves guarantee a profit, since other factors such as size of business and labor efficiency may limit or partly offset the advantage of higher rates of production.

## FUNDAMENTALS OF FARMING

TABLE 8

RELATION OF BUTTERFAT AND MILK PRODUCTION PER COW AND LABOR INCOME FOR FIVE LOWEST AND FIVE HIGHEST PRODUCING HERDS IN STUDY, 1940

Average per Farm per Year

|  | Five Low Producing Herds | Five High Producing Herds |
|---|---|---|
| Butterfat Production per Cow—Lbs. | 207 | 363 |
| Milk Production per Cow—Lbs. | 5479 | 9113 |
| Labor Income | $386 | $1,293 |
| Number of Cows | 31 | 26 |
| Total Farm Receipts | $6,838 | $8,431 |
| Total Capital | $20,242 | $21,134 |

### LABOR EFFICIENCY

Man labor is an important item in cash expenditure of farm production. On dairy farms in New Jersey, around 25 to 30 per cent of the cash expenses are for man labor, and on fruit and

FIGURE 3

INDEX NUMBERS OF HIRED FARM LABOR AND FARM PRICES IN NEW JERSEY (*1910–14=100*)

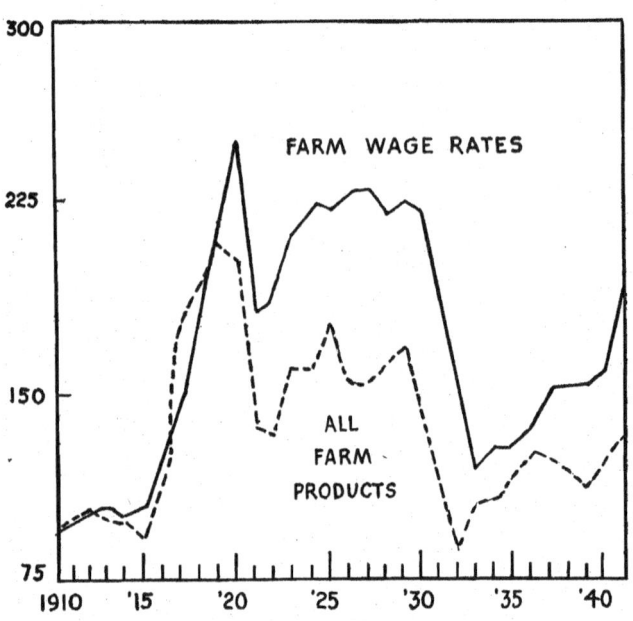

## FARM MANAGEMENT

vegetable farms the labor expenses represent 30 to 40 per cent or even more of the total farm expenses, depending upon the type and intensity of the farm set-up. Another factor in the farm

*Cutting Lima Beans, Using Horses.*

labor situation is that of wage rates, which, since the last war, have tended to be relatively higher than prices obtained for farm products, as based upon the ratios obtaining in 1910.

*Cutting Lima Beans, Using Machinery: power machinery for producing and harvesting reduces man-labor requirements to an average of less than 35 hours per acre.*

Efficiency in the use of man labor ranks in importance with size of business and rates of crop and livestock production in determining the success of the farm as a business enterprise. In recent years, farmers throughout the nation have more generally changed from horses to tractors and other mechanized implements for the production of farm products. The use of these machines has greatly reduced the labor necessary; however, the reduction has not been so great as that found in industry, since there are many farm operations for which machinery has not yet been developed.

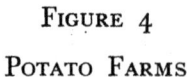

FIGURE 4

POTATO FARMS

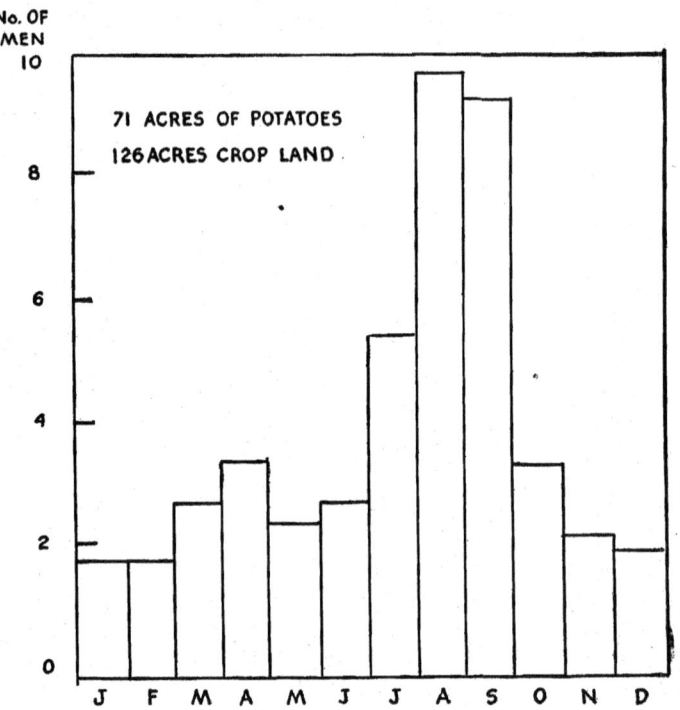

Some idea of the development of the application of machinery to agriculture is shown in the case of central New Jersey potato farms. In 1926, before extensive motorization began, surveys showed that 37 hours of man labor, 50 hours of horse labor, and 2.1 hours of tractor use were required in the production per acre of potatoes up to harvest. Ten years later (1936), a study of 10 farms on which general-purpose tractors were used for most of the field operations, an average of 17 hours of man labor, 6.8 hours of tractor use, and 4.1 miles of truck use were required for

FIGURE 5

MARKET GARDEN FARMS

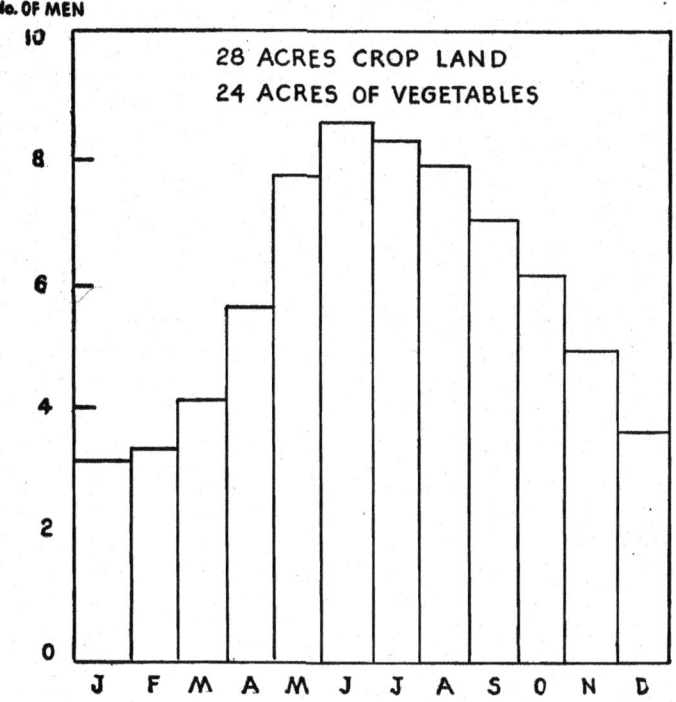

FIGURE 6

DAIRY FARMS

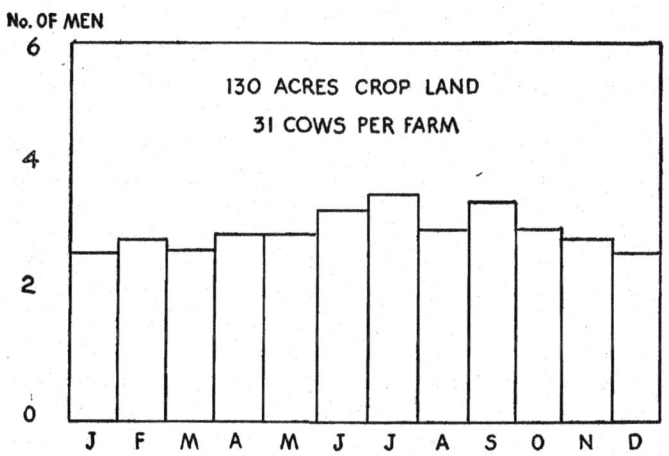

## FUNDAMENTALS OF FARMING

the production per acre of potatoes up to harvest. This was a saving of about 2 days of man labor per acre during the growing season. Some of this saving in labor and horse use, however, was offset by considerably increased investment in machinery. This general trend to mechanized farming has placed even greater emphasis on the need for skill and efficiency in use of labor. In many instances, it has widened the difference in labor income between large- and small-size farms or business volume.

Another part of the labor problem on farms is the effect of type of farm or farming enterprise upon the distribution of employment throughout the year. The rather marked seasonal variation in labor requirements for specialized potato farms is shown in Figure 4. The more regular seasonal increase and decrease in labor requirements on market-garden-vegetable farms is shown in Figure 5. Specialized dairy farms growing only feed crops have a relatively level man-labor requirement throughout the year, as shown in Figure 6. These charts show the average number of men, including the operator, employed month by month on representative farms selected for study. In each figure, the average number of acres per farm devoted to the principal crop (or the average number of producing animals, in the case of dairy farms) and the average total acreage of crop land is indicated.

There is considerable variation in the ability of farm operators to organize and plan their business and manage the labor. An illustration is provided in the case of two southern New Jersey vegetable farms of approximately the same size, type, and intensity of crop production, and which had about the same crop-yield index. One farm was spending about 30 cents of every dollar for man labor and obtained a labor income of $1,535; the other farm was spending 41 cents of every dollar received for man labor and had a labor income of minus $816. Both of these farms were of a favorable size and obtained satisfactory crop yields. However, the one farm was very inefficient in use of man labor, to the extent that there was no return left to the operator for his work.

The general principle of the higher labor incomes resulting from a higher degree of labor efficiency is illustrated in a study of 1525 Michigan farms (Table 9). The farms with more productive work days per man per year on the average lost less

## FARM MANAGEMENT

money during the period of falling prices and obtained a higher labor income during the period of rising prices.

TABLE 9

RELATIONSHIP OF LABOR EFFICIENCY TO LABOR INCOME ACCORDING TO 1,525 FARM RECORDS IN CENTRAL MICHIGAN DURING FALLING AND RISING PRICE PERIODS[1]

| Productive work days per man per year | Labor Income Falling Prices (1930–32) | Rising Prices (1933–38) |
|---|---|---|
| Less than 175 | –$946 | $ 256 |
| 175–224 | –737 | 540 |
| 225–274 | –697 | 746 |
| 275–324 | –679 | 1,065 |
| 325 or more | –593 | 1,518 |

[1]Michigan State College and Agricultural Experiment Station, *Circ. Bul. 182*, by Doneth and Wright.

Every farm operator needs to give careful consideration to the details of his farm organization, use of machinery, the layout of fields, arrangement of buildings, and diversity of enterprises to provide for the most productive employment of skilled labor.

### FARM PRICES AND LABOR INCOME

The farmer has some degree of control over the set-up of his business in respect to size, rates of production, efficiency of labor, and other factors involved in the management of his farm. But he has no control over the general trend of prices. However, he should be acquainted with the trends and movement of prices as they affect his business, so that he may make adjustments which would increase his profits from the farm. An illustration of the general effect of the price level on the income for different sizes of farms is shown for Indiana farms for the period 1929–38 (Figure 7). It will be noted that in two of the depression years, 1931 and 1932, net farm returns were very much in the "red" and that the larger farms showed a greater loss than the small farms. In the period of higher prices, 1935–38, labor incomes were considerably higher and the larger farms had a distinct advantage.

For some farm products there are very marked cycles or periods of high and low prices, as in the case of hogs and cattle. There is

## FUNDAMENTALS OF FARMING

also a variation in the long-time price relationship between different agricultural products. This is illustrated by the 1910-42 price ratios of vegetables produced for fresh market in New Jersey and all farm products in New Jersey (Figure 8).

*Regular servicing of equipment is an important part of good farm management.*

It will be noted that in the period from 1920 to 1929 the prices for vegetables averaged considerably higher (based on the 1910-14 ratios) than the average price of all farm products in New Jersey, whereas in the period 1933 to 1940 the prices of vegetables were considerably lower than the average price for all farm products. In other words, vegetable growing was relatively profitable in the '20's and generally unprofitable in the '30's, due

FIGURE 7

RELATION OF SIZE OF FARM TO LABOR INCOME IN YEARS WITH DIFFERENT LEVELS OF FARM PRICES[1]

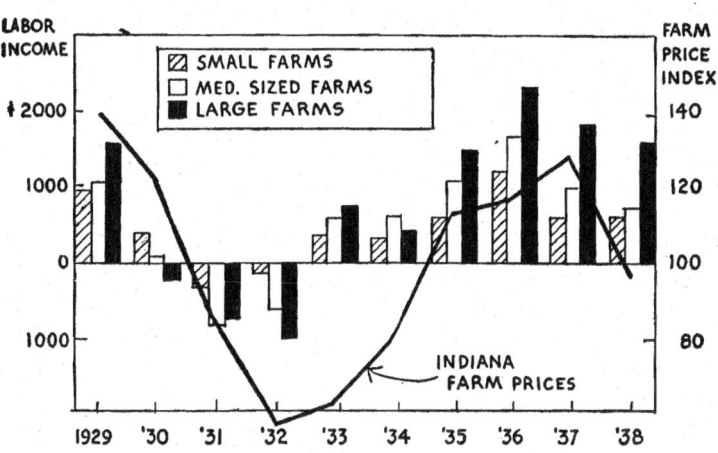

[1]Indiana Agri. Exper. Station, *Bul. 452*, by Robertson.

FIGURE 8

INDEX OF PRICES RECEIVED BY NEW JERSEY FARMERS FOR VEGETABLES FOR MARKET AND ALL FARM PRODUCTS
*1910–14=100*

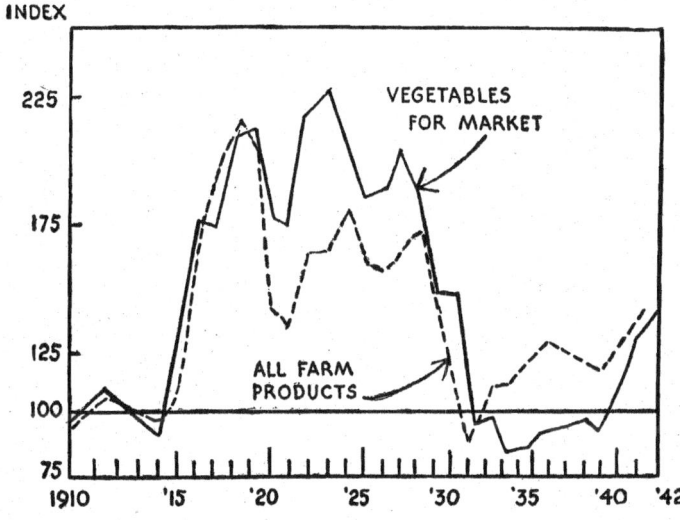

## FUNDAMENTALS OF FARMING

to long-time changes in price relationships over which the individual producer had no control.

Farm earnings are also influenced by the trends in general relation between prices received for farm products and prices

FIGURE 9

INDEX OF PRICES RECEIVED BY FARMERS AND FARM WAGE RATES IN THE UNITED STATES
*1910–14=100*

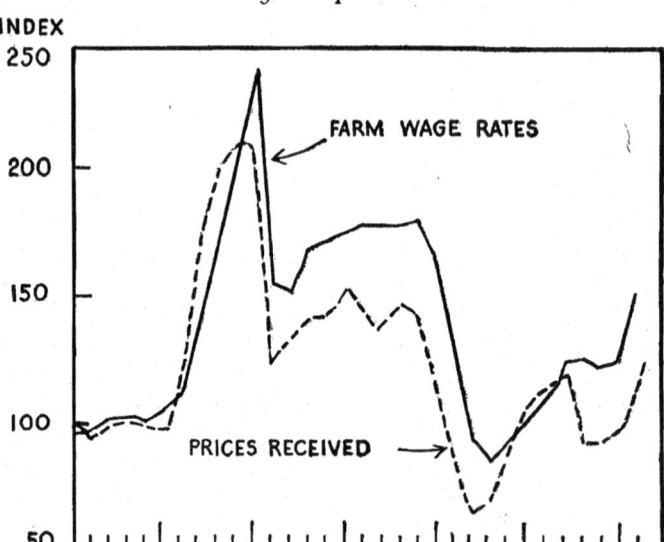

paid for items entering into the cost of production. As illustrated in Figure 9, farm prices received in the United States fell after the last war to a lower level than the price paid for hired labor—an important item in the cost of farm operation. The unfavorable relation existing between prices received and prices paid in the period of the '20's and early '30's directed a great deal of thought to proposals for farm relief and resulted in Congressional action designed to benefit the farmer.

Many farm products have a marked seasonal variation during the year, as illustrated for spinach in Figure 10, and in some instances farmers can plan their production so the larger part of their marketing will occur when seasonal prices are relatively

## FARM MANAGEMENT

more favorable. There are, of course, limitations to the adjustments which can be made along this line. However, the records show that the more successful farmers obtained higher prices for

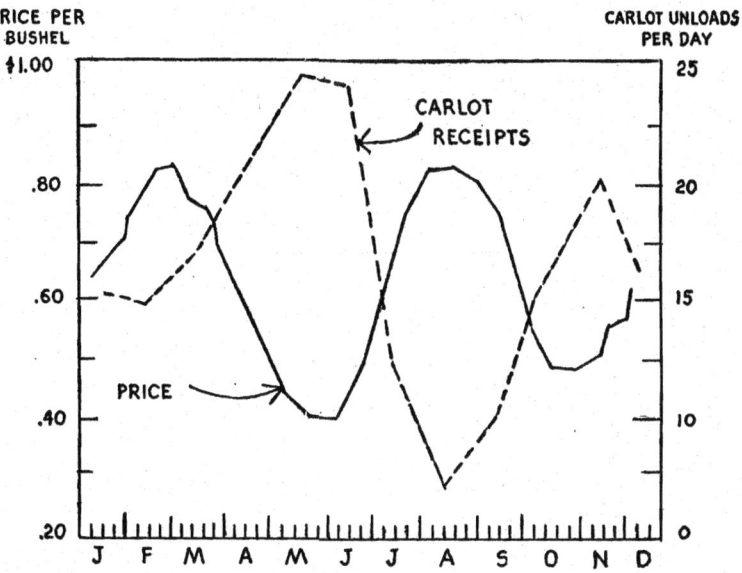

FIGURE 10

SEASONAL PRICE OF SPINACH AND THE CAR-LOT RECEIPTS ON THE NEW YORK WHOLESALE MARKET
(*5-year average 1933–37*)

their products due, in part, to their study of seasonal price trends. The more a farmer knows about the movement of prices, the better is he able to judge the relative profitableness of possible enterprises adapted to his farming conditions.

### COMBINATION OF ENTERPRISES AND THE FARM BUDGET

In any area, over a period of time, certain enterprises tend to be more profitable than others. The aim of the farm manager naturally should be to combine the more profitable enterprises in the set-up of his farm organization. The keeping of a complete set of cost accounts would be very helpful in determining the most profitable enterprises. It is necessary, however, to keep in mind that the main objective is the largest net income from

## FUNDAMENTALS OF FARMING

the farm business as a whole, and, from that point of view, the farm budget method provides a simpler approach to the determination of alternative combinations of enterprises.

The general method of determining the relative advantages of different combinations of enterprises, applicable to any farm business, is illustrated for a 140-acre dairy and cash-crop farm in Table 10. The operator can put down on paper the present

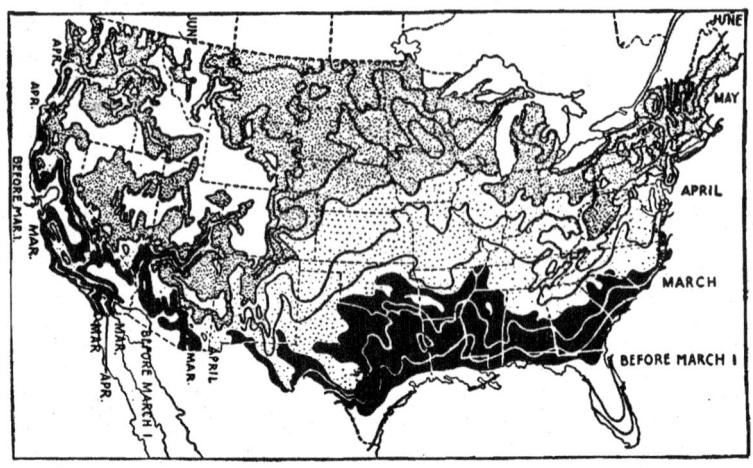

*In considering alternative combinations of crops, the farmer must allow for possible differences in crop hardiness under his particular climatic conditions. This outline map of the United States, showing the dates of the average occurrence of the last spring frost (U.S.D.A.) will help him to plan a successful combination of enterprises.*

set-up of his farm business, as indicated in the table, including receipts and expenses. In the illustrative table, the business consists primarily of a dairy herd of 14 cows, the growing of roughage feed crops for the cows, together with the production of 70 acres of Lima beans and 20 acres of peas for processing. Lima beans and peas are relatively extensive vegetable crops in their demands for soil area, and the question might be raised as to the economic advantage or disadvantage of converting part of the acreage from these crops to more intensive vegetable crops, such as asparagus and tomatoes for processing. The operator can estimate, by observation of neighbors or consultation with the County Agent, the yields which he might anticipate and the average or anticipated prices, and so calculate the change in income which would result from the change of crops. He can

# FARM MANAGEMENT

Table 10

## COMPARISON OF PRESENT AND ALTERNATIVE ORGANIZATION FOR A 140-ACRE DAIRY AND CASH-CROP FARM[1]

|  | Yield per Acre | Price Received per Unit | Present Plan | Alternative Plan 1[2] |
|---|---|---|---|---|
| CROPS: |  | Dollars | Acres | Acres |
| Lima Beans (freezer) | 1375 lbs. | .04 | 70 | 50 |
| Peas (freezer) | 670 lbs. | .04 | 20 | 20 |
| Asparagus (freezer) | 2500 lbs. | .05 | — | 10 |
| Tomatoes (canhouse) | 5.1 tons | 15.04 | — | 10 |
| Corn for Grain | 25 bu. | — | 4 | 4 |
| Corn for Silage | 10 tons | — | 9 | 9 |
| Alfalfa | 3.7 tons | — | 8 | 8 |
| Crimson Clover and Turnips |  | — | 8 | 8 |
| Total Crop Sales |  |  | $4,380 | $5,326 |
| LIVESTOCK: |  |  | Number | Number |
| Cows |  |  | 14 | 14 |
| Heifers |  |  | 6 | 6 |
| Bull |  |  | 1 | 1 |
| Horses |  |  | 3 | 3 |
| Chickens |  |  | 75 | 75 |
| Milk Sales—1126 cwt. |  |  | $2,371 | $2,371 |
| Total Livestock Sales |  |  | 2,877 | 2,877 |
| Miscellaneous Receipts |  |  | 863 | 863 |
| Total Gross Receipts |  |  | $8,120 | $9,066 |
| EXPENSES: |  |  |  |  |
| Man Labor |  |  | $1,470 | $1,845 |
| Fertilizer |  |  | 1,277 | 1,349 |
| Seeds and Plants |  |  | 1,077 | 975 |
| Feed |  |  | 505 | 505 |
| Fuel and Oil |  |  | 350 | 359 |
| Containers |  |  | — | 20 |
| Taxes |  |  | 324 | 324 |
| Building and Equipment Upkeep |  |  | 1,020 | 1,020 |
| Other Expenses |  |  | 637 | 765 |
| Interest on Investment |  |  | 1,006 | 1,006 |
| Total Expenses Including Interest on Investment |  |  | $7,892 | $8,168 |
| Operator's Labor Income |  |  | $228 | $898 |

[1]Agricultural Economics Report No. 55, "Farm Management Problems and Costs and Returns in Producing Lima Beans and Peas for Quick Freezing, Cumberland County, New Jersey," New Jersey State Agricultural College (mimeo.), in cooperation with the U. S. Department of Agriculture.

[2]Includes asparagus for freezing and tomatoes for canhouse.

also figure, item by item, the change in expenses which would result from the substitution of these more intensive crops for those which are more extensive in their demands for growing area.

The alternative combination of enterprises shown above resulted in an increase in gross receipts of $946. This additional gross income was obtained with an additional expenditure of only $276, resulting in a net gain to the operator of $670 in his labor

*Inter-cropping of spinach between rows of celery provides for an intensive use of land.*

income. Any operator, by similar analyses, can calculate other alternative combinations of enterprises for his own farming conditions.

In setting up the farm budget or considering alternative plans, the operator should keep in mind the principles which have already been discussed, namely, the size of business, practices which will increase yields, the efficiency in use of labor, including the distribution of labor requirements for different crops through the year, and the trend of price relationships as they affect his business. In addition, consideration should be given to the rotation of crops, the maintenance of fertility and control of soil erosion, the investment necessary for machinery or buildings for different enterprises, the matter of risk which may be increased with extreme specialization, and distribution of income through-

## FARM MANAGEMENT

out the year. Studies involving large numbers of farms show that the better a farm manager exceeds the community average in his control of efficiency factors, the greater will be his labor income. Farming is carried on under changing conditions of supply, demand, and price levels, and the successful farmer is continually on the alert to adjust his farm organization and operation to meet these changes.

### Farm Records and Accounts

Farm records can be of value to the farm operator in many ways. There is, however, no one simple system which will answer all questions. Accounting methods used in the business world are usually not well adapted to the farming business. The system which is of most value to the farmer is the one which will show him the most about his business without making too great demands upon his time. A farm inventory provides basic information which far out-values the time required for taking it. Any farmer, with his knowledge of market values as observed at farm sales, can place reasonable values on the different items of equipment, livestock, and farm property which he owns. The inventory should be taken about the same time each year; on many farms, January 1st is a convenient time, although the date may vary for different types of farms.

The State Colleges of Agriculture or the County Agricultural Agents can recommend the type of inventory record book best suited to the individual's needs. An actual farm inventory of a New Jersey dairy and poultry farm (Table 11) illustrates the type of information which should be included and indicates the general procedure to be followed in taking inventory of any type of farm. This record shows the value of each item of capital investment and the total investment in the farm business at the beginning and end of the farm year. The total figures will indicate whether there was an increase or decrease in total inventory. Such an inventory of the internal farm business can be particularly useful for making a credit statement or for figuring the operator's labor income.

In making a credit statement, such items as notes and accounts receivable, cash on hand, mortgages, and notes and accounts payable would be added to the inventory as shown in Table 12.

# FUNDAMENTALS OF FARMING

Table 11

## A FARM INVENTORY

|  | Value—Beginning of Year | |
|---|---|---|
|  | 1939 | 1940 |
| **REAL ESTATE—** | | |
| Dwelling House | $3,500 | $3,500 |
| Other Buildings | 3,250 | 3,250 |
| Land | 6,750 | 6,750 |
| Total | 13,500 | 13,500 |
| **EQUIPMENT—** | | |
| Tractor | $250 | $250 |
| Truck | 100 | 250 |
| Disc Harrow | 100 | 90 |
| Corn Harvester | 120 | 115 |
| Grain Binder | 90 | 85 |
| Hay Rake | 45 | 40 |
| Wood Saw | 10 | 10 |
| Harness | 20 | 15 |
| .... | .... | .... |
| .... | .... | .... |
| .... | .... | .... |
| .... | .... | .... |
| Total | $3,032 | $3,295 |
| **LIVESTOCK—(Subdivided by kind)** | | |
| Horses—Nellie | $125 | $100 |
| Maud | 175 | 150 |
| Total | 300 | 250 |

|  | No. | Value | No. | Value |
|---|---|---|---|---|
| Poultry—Hens | 285 | $350 | 424 | $530 |
| Cockerels | 600 | 350 | 36 | 45 |
| Total |  | $700 |  | $575 |

| | | |
|---|---|---|
| **CATTLE—COWS** | | |
| (By name or number) | | |
| Alice | $110 | $100 |
| Helen | 90 | 85 |
| Brownie | 125 | 100 |
| .... | .... | .... |
| .... | .... | .... |
| .... | .... | .... |
| .... | .... | .... |
|  | Value—Beginning of Year | |

## FARM MANAGEMENT

LIVESTOCK (Continued)      *1939*     *1940*

CATTLE—HEIFERS
(By name or number)

|  | 1939 | 1940 |
|---|---|---|
| Spot | $ 35 | $ 75 |
| Blackie | 10 | 40 |
| .... | .... | .... |
| .... | .... | .... |
| .... | .... | .... |
| .... | .... | .... |
| Calves |  | 15 |
| Bulls—Tom | 125 | 125 |
| .... | .... | .... |
| .... | .... | .... |
| Total | $2,585 | $2,825 |

CROPS AND SUPPLIES—     *1939*           *1940*

|  | Amt. | Price | Total Value | Amt. | Price | Total Value |
|---|---|---|---|---|---|---|
| Hay, ton | 25 | $14.00 | $350.00 | 34 | $20.00 | $680.00 |
| Corn, bu. | 1650 | .30 | 495.00 | 1500 | .35 | 525.00 |
| .... | .... | .... | .... | .... | .... | .... |
| .... | .... | .... | .... | .... | .... | .... |
| .... | .... | .... | .... | .... | .... | .... |
| Mixed Mill Feeds, pounds | — | — | — | 1500 |  | 16.50 |
| Fertilizer, pounds |  |  |  | 500 |  | 12.00 |
| Crates |  |  |  | 2 |  | 2.00 |
| .... |  |  |  | .... |  | .... |
| .... |  |  |  | .... |  | .... |
| .... |  |  |  | .... |  | .... |
| Total |  |  | $1,417.75 |  |  | $1,900.00 |

### SUMMARY—INVENTORY

|  | Value 1939 | Value 1940 |
|---|---|---|
| Real Estate | $13,500.00 | $13,500.00 |
| Equipment | 3,032.00 | 3,295.00 |
| Horses | 300.00 | 250.00 |
| Poultry | 700.00 | 575.00 |
| Cattle | 2,585.00 | 2,825.00 |
| Crops and Supplies | 1,417.75 | 1,900.00 |
| Total | $21,534.75 | $22,345.00 |

Change in Inventory—Increase $810.25.

## FUNDAMENTALS OF FARMING

Such a statement shows the present worth of the individual for two periods of time and indicates whether or not he is better off financially at the end of the year. It should be observed that this type of statement records not only the internal phase of the farm investment, but also the status of the external financial transactions included under notes and accounts receivable and payable. This kind of statement is very helpful in applying for credit.

TABLE 12

A CREDIT STATEMENT

|  | Inventory | |
|---|---|---|
|  | Beg. of Year 1939 | Beg. of Year 1940 |
| **ASSETS—** | | |
| Real Estate | $13,500.00 | $13,500.00 |
| Equipment | 3,032.00 | 3,295.00 |
| Horses | 300.00 | 250.00 |
| Poultry | 700.00 | 575.00 |
| Cattle | 2,585.00 | 2,825.00 |
| Crops and Supplies | 1,417.75 | 1,900.00 |
| Notes and Accounts Receivable | 1,325.00 | 1,365.00 |
| Cash on Hand | 221.00 | 133.00 |
| Total | $23,080.75 | $23,843.00 |
| **LIABILITIES—** | | |
| Mortgages | $ 7,690.00 | $ 7,390.00 |
| Notes and Accounts Payable | 2,365.00 | 2,325.00 |
| Total | $10,055.00 | $ 9,715.00 |
| **PRESENT WORTH** | $13,025.75 | $14,128.00 |

To study more fully the details of the internal phase of the farm operation for the year, the simplest system is a farm cash account. This can take the form of a simple recording each day of the financial transactions or sales and purchases, as illustrated in Table 13. Items of personal expense for family living should not be included in this account. Furthermore, since this is a study of the internal economy of the farm, a separate recording of credit transactions should be kept. In other words, if two tons of dairy feed were purchased on credit, the total cost should be entered in the cash paid column and a special record maintained of the purchases and payments on credit. Thus, at the end of the year, the gain or loss on a receipts-and-expense basis can be determined. This result can then be added to the amount representing change in inventory, as shown in Table 13, and the resulting figure will represent the farm income. Interest on the average capital for the year can be computed at the current rate, and this amount subtracted from the farm income will give the labor income.

*Keeping accurate records reveals costs of production and helps in planning future operations.*

Labor income is the return to the operator for his year's work on the farm; in addition, he has the use of the house and products furnished by the farm, for family living. This is a fair measure of his success in the operation of the farm for the year. At the end of the year, or at the end of each month, the individual items entered in the cash account may be classed under appropriate headings, depending on the type of farm, such as milk sold, eggs sold, labor expense, poultry-feed expense, dairy-feed expense, fertilizer, and containers. Some farm account books have provided for this classification as an extension of the cash-received-and-paid entry. This classification of the items is valuable in summarizing the farm business for the year and is particularly useful in preparing income tax forms.

## FUNDAMENTALS OF FARMING

TABLE 13

A FARM CASH ACCOUNT

|  |  |  |  | Cash | |
|---|---|---|---|---|---|
| Date | Amount | Description | Price | Received | Paid |
| Jan. 3 |  | Lumber |  |  | $ 4.50 |
| " 5 | 2 cases | Eggs | $ .21 | $ 12.56 |  |
| " 6 | 2000 lbs. | Poultry Feed | 1.91 |  | 38.25 |
| " 10 | 3 cases | Eggs | .20 | 18.08 |  |
| " 11 | 2300 lbs. | Dairy Feed | 1.64 |  | 37.65 |
| " 12 |  | Milking Machine Rubbers |  |  | 6.00 |
| " 12 | 150 | Broilers |  | 103.86 |  |
| " 13 | 25 gal. | Gasoline | .125 |  | 3.13 |
| " 14 | 11,241 lbs. | Dec. Milk, 4.6% B.F. | 3.71 | 416.95 |  |
| " 14 |  | Veterinary |  |  | 5.00 |
| " 14 | 1 | Calf |  | 2.00 |  |
| " 15 | 2700 lbs. | Poultry Feed | 1.93 |  | 52.15 |
|  |  | .... |  | .... | .... |
|  |  | .... |  | .... | .... |
|  | Total for Year |  |  | $8,054.57 | $6,929.14 |
|  | Cash Farm Receipts Less Cash Expenses |  |  | $1,125.43 |  |
|  | Change in Inventory—Increase |  |  | 810.25 |  |
|  | Farm Income |  |  | 1,935.68 |  |
|  | Interest on Average Capital ($21,939.88)@5% |  |  | 1,096.89 |  |
|  | Labor Income |  |  | 838.79 |  |

A single-enterprise account is very helpful in studying the details of the economy of one or more of the important phases of the farm business. Such a record involves a complete listing of all of the items of expense in connection with the production of that enterprise and a listing of the credits to the enterprise. Care should be taken to make the record complete and to include all of the items involved. The expenses, for example, should include not only the direct-cash expenses (as for seed, fertilizer, containers, etc., in the case of the crop), but also a charge for use of land, use of all machinery, all hours of man labor, etc. Similarly, the credits (in the case of a crop) might include, in addition to receipts for what was sold, the worth of the portion fed on the farm or used by the family or saved for seed. Enterprise accounts can be of most value in studying the farm business in an effort to make the enterprise more profitable, although they are also valuable as information on the cost of production. Most State Agricultural Colleges have available single-enterprise

## FARM MANAGEMENT

farm account books which can be used for keeping a record of the individual enterprises. These colleges usually cooperate with farmers who keep single-enterprise accounts and publish summaries of these which are of value to the individual for comparison with his own enterprise accounts.

*In some areas, producers share the cost of employing a competent bookkeeper who, at periodic intervals, assists participating farmers in keeping and analyzing their farm accounts.*

A complete farm-cost set-up is a more all-inclusive system than the one described above, but is so complicated and takes so much time that it is not practical for most farmers unless they have outside supervision in keeping them and help in summarizing them.

In several states, farm account associations have been set up by farmers to obtain help in the keeping and analyzing of farm records. This is usually on a cooperative basis between the farmers' accounting organization and the State Colleges. The cooperating farmers pay a fee, either a fixed amount or an amount varying with the size of the farm. This provides for the employment of a field man who has had training and experience in farm-management problems. He goes from farm to farm and

# FUNDAMENTALS OF FARMING

assists the operator with his record-keeping problems and discusses with the farmer the interpretation of the records kept.

The compilation of the inventory and cash account provides the operator with a basis for making a credit statement and studying his income and his business with others, as regards volume of business, rates of production, labor efficiency, etc. Many farmers make a good start at the beginning of the year, but gradually get behind in their accounts, with the result that they are not completed. Farm accounts are a worthwhile part of every business and every efficient farmer will keep his records complete and up-to-date.

## Suggested Readings

AGRICULTURAL PRICE ANALYSIS, by Geoffrey S. Shepherd. (Iowa State College Press, Ames, Iowa.)

AGRICULTURAL SITUATION. (Monthly Publication of Bureau of Agricultural Economics, U. S. Department of Agriculture, Washington, D. C.)

DAIRY FARM MANAGEMENT AND COSTS IN PENNSYLVANIA, by W. L. Barr. (*Bulletin 421*, Pennsylvania Agricultural Experiment Station, State College, Pa.)

ECONOMIC STUDIES OF POULTRY FARMING IN NEW YORK, by E. G. Misner and A. T. M. Lee. (*Bulletin 684*, Cornell University Agricultural Experiment Station, Ithaca, N. Y.)

EFFECT OF CHANGES IN MILK AND FEED PRICES AND IN OTHER FACTORS UPON MILK PRODUCTION IN NEW YORK, by Merton S. Parsons. (*Bulletin 688*, Cornell University Agricultural Experiment Station, Ithaca, N. Y.)

ELEMENTS OF FARM MANAGEMENT, by John A. Hopkins. (Prentice-Hall, Inc., New York.)

FACTORS CAUSING VARIATIONS IN EARNINGS AMONG DAIRY FARMERS IN SOUTHWESTERN MINNESOTA, by G. A. Pond, W. P. Ranney, and C. W. Crickman. (*Bulletin 314*, Minnesota Agricultural Experiment Station, University Farm, St. Paul, Minn.)

FARM ECONOMICS, MANAGEMENT AND DISTRIBUTION, by Frank App and Allen G. Waller. (J. B. Lippincott Company, Philadelphia.)

FARM MANAGEMENT ON CENTRAL MAINE FARMS WITH DAIRY ENTERPRISES, by Emil Rauchenstein and A. E. Watson. (*Bulletin 408*, Maine Agricultural Experiment Station, Orono, Maine.)

FARM MANAGEMENT AND MARKETING, by V. B. Hart, M. C. Bond, and L. C. Cunningham. (John Wiley and Sons, New York.)

FARM ORGANIZATION AND MANAGEMENT, by G. W. Forster. (Prentice-Hall, Inc., New York.)

FARM PRACTICES AND THEIR EFFECTS ON FARM EARNINGS, by M. L. Mosher and H. C. M. Case. (*Bulletin 444*, Illinois Agricultural Experiment Station, Urbana, Ill.)

# MARKETING

## L. A. BEVAN
## W. R. STONE

*Prior to his appointment as director of the New Jersey Extension Service in 1939, Professor Laurence A. Bevan had devoted many years of study and work to problems of marketing and distributing farm products. He is a former director of the Bureau of Markets in his home state of Massachusetts, a one-time agricultural agent for the Boston Chamber of Commerce, and for four years was Extension Service economist in marketing at Rutgers. He has been a farm hand, County Agent, and teacher of vocational agriculture. At the request of the U. S. Department of Agriculture, he has collaborated in studies, and contributed to plans for improvement of terminal produce markets in New York and Philadelphia. It was at the request of the same Federal Agency that he was granted a three months' leave, in 1940, to survey farm marketing problems that would confront the huge area to be irrigated in the Columbia River Basin in Washington. This assignment called for recommendations involving products to be grown on a million-and-a-quarter acres of land. Professor Bevan is a past president of both the New England Association of Marketing Officials and the National Association of Marketing Officials. He is a graduate of Massachusetts State College.*

*A native Jerseyman, trained in horticulture at Rutgers, and the dean in point of service among New Jersey's County Agricultural Agents, W. Raymond Stone fathered a unique system of accredited farm roadside markets in his home county of Bergen. That densely populated area, situated across the Hudson River from New York City, still maintains a flourishing agricultural industry despite industrial and residential developments. This is due in no small measure to the effectiveness with which Bergen farmers sell direct to consumers at the roadside. Their early and at first promising efforts in this direction were hampered by unscrupulous hucksters who, posing as farmers, sold inferior produce to unsuspecting consumers. County Agent Stone proposed a plan for farm market accreditation, with the Bergen County Chamber of Commerce serving as an impartial accrediting agency. Acceptance of the proposal enabled consumers to identify markets of bona fide farmers pledged to sell only high-quality, honestly-graded products.*

*The plan has since been copied widely in other states. It represents only one of the means by which Mr. Stone has aided his farmers in their efforts to sell efficiently, with benefit to producer and consumer alike, at the roadside.*

# ECONOMICS OF MARKETING

## L. A. Bevan

SATISFACTION to a farmer in marketing his products usually means that his goods were in good demand, were attractive to the buyers, and commanded what the farmer believes to be a good price. But this first sale of goods by the farmer is usually only the first step in the marketing process, which is completed only when the products are in the hands of the final consumer. When food products need much processing or are highly perishable or are grown far from the consuming point, marketing becomes an involved and often costly procedure.

Marketing has a close relation to the growing of crops or livestock, since the needs of consumers or the demands of buyers as to the type or quality of the products must be kept in mind. For instance, McIntosh apples have increased in demand in the Northeast, taking the place of older varieties. In the growing of turkeys, a broad-breasted strain is being developed which is meeting with good favor by consumers. The vegetable, broccoli, is now grown and supplied in considerable quantities to many markets, but it was not well known only a few years ago. Marketing operations, as such, however, begin only when the product is ready to be moved from the farm. Products must be put up in such a manner that they will interest buyers and in condition to be transported. A buyer must be contacted or the goods delivered to a market place.

The farmer has some choice in the matter of (1) selling immediately after harvest or storing for later sale, (2) selling to a local or a distant market, and (3) deciding whether his market outlet shall be a country buyer, a commission firm, a farmers' cooperative, or, if he is favorably situated, selling direct to the consumer. Within these possible markets, however, the farmer has little influence upon the prices he may expect for his products.

## MARKETING

Production, packing, and choice of market by the individual farmer are only the preliminary steps in the long and often difficult process of getting food products from the several million farms all over the United States to the 100-odd-million consumers. The bulk of farm products is raised at a considerable distance from consuming centers, and individual farmers grow

*Candling Eggs—an Important Step in Safeguarding Quality*

relatively large quantities of a limited number of products, while a single family demands small-unit amounts of a wide range of commodities. Furthermore, few farm products can be consumed as they come from the fields, and much packaging and processing are necessary before they are ready for the ultimate user.

Individual families are not interested in buying food products as they are harvested from the field or in the wholesale lots usually put up at the farm. They do not buy a 40-quart can of milk, a 100-pound sack of potatoes, or a 30-dozen crate of eggs.

## PRODUCING FOR PROFIT

Eggs are candled and packed in cartons, milk is pasteurized and bottled, poultry picked, stored, and dressed, potatoes graded and done up in small packages. In large cities today, apartment dwellers demand much service. They make frequent purchases

*Ineffective packing in this railroad car was responsible for breakage and loss.*

of small amounts at a time, such as a half dozen or even 3 eggs, half a squash, 3 pounds of potatoes, or a quarter of a pound of butter.

This in part explains why there is a wide difference between the retail price of food products paid by consumers and the price received by farmers. The difference is represented by marketing costs and is called spread or margin. Analysis of this spread or margin between farm and retail prices shows it to vary with a

## MARKETING

number of factors, among which are: (1) the type of the product, (2) the distance from market, (3) the amount of handling or processing necessary, (4) the amount of waste and spoilage involved, and (5) the bulkiness of the product. Margins will also vary over a several years' period, depending upon price levels, improvements in marketing methods, and changes in wage rates.

A report of the U. S. Department of Agriculture[1] on the spread of 58 food products shows that just before the first World War farmers received 53 per cent of the retail value; during the early part of the depression, 36 per cent; and in 1936 and 1937, 45 per cent. This report concludes that there is a close relationship between hourly wage rates and charges for transportation, processing, and other marketing costs. If, therefore, margins are high, farmers located close to consumers can often increase their income by performing much of the marketing services.

Marketing functions are often classified as (1) assembling, (2) processing or packaging, (3) transporting, (4) financing, (5) risk taking, (6) selling or exchanging ownership, and (7) distributing or dispersing. Some of these functions are more costly and involve more risk than others. An illustration[2] will show this situation in terms of the distribution of the retail food dollar for two country-wide products:

|                      | Fresh Fruits and Vegetables | Dairy Products |
|----------------------|---------------------------|----------------|
| Farmer Received      | 29.4%                     | 47.1%          |
| Broker               | 1.5                       | .6             |
| Transportation Agency| 20.2                      | 5.0            |
| Processor or Packer  | 13.6                      | 27.6           |
| Wholesaler           | 4.3                       | 5.8            |
| Retailer             | 31.0                      | 13.9           |
|                      | 100.0%                    | 100.0%         |

These figures, based on analyses over a wide area, show where costs are occasioned for various services necessary to get food to consumers. It can be seen that transportation was 4 times as high for fruits and vegetables as for dairy products, and that the

[1] Been and Waugh, "Price Spreads between the Farmer & the Consumer," Bureau of Agricultural Economics, U. S. Department of Agriculture.
[2] Lazo and Bletz, "Who Gets Your Food Dollar?"

cost of retailing was more than twice as great for the fruits and vegetables. In general, it can be said that those products which require transportation from long distances (lettuce or carrots from California), those which have a high spoilage rate (berries), and those that need considerable processing (canned goods), have relatively high marketing costs.

New or different methods of transportation, processing, or merchandising will change the picture, although no way has yet been devised to do away with the agencies that are engaged in the marketing functions. Farmers as well as others believe that marketing costs are too high and that improvements in the marketing systems would result in a substantial increase in the prices they would receive for their products. While changes are sure to come, some of which will result in more efficient marketing methods, all of the savings will not be reflected in returns to farmers, but part will likely go to consumers.

Selling farm products at what may be considered good prices may have little relation to sound marketing methods, because the general price level may be the controlling factor. In one season, a farmer may believe he or his marketing agency has performed good service, since his returns were satisfactory. In another season, with products of similar quality, returns may be considerably lower due to a general fall in demand or a much lower price level. In the latter situation, farmers are liable to blame the marketing system, but—whatever its shortcomings—it should not be held accountable for variations in price levels. Yearly variations in prices received by farmers from 1910 to 1942 for farm products as a whole, as well as for dairy products, vegetables, eggs, and chickens, are shown in the following table:

## MARKETING

### PRICES RECEIVED BY FARMERS, 1910-42*

| | All Farm Products | Dairy Products | 17 Vegetables for Fresh Market Shipment | Eggs | Chickens |
|---|---|---|---|---|---|
| | (1910-14 =100) | (1922-41 =100) | (1924-29 =100) | (1910-14 =100) | (1910-14 =100) |
| 1910 | 102 | — | — | — | — |
| 1911 | 95 | — | — | — | — |
| 1912 | 100 | — | — | — | — |
| 1913 | 101 | — | — | — | — |
| 1914 | 101 | — | — | — | — |
| 1915 | 98 | — | — | — | — |
| 1916 | 118 | — | — | — | — |
| 1917 | 175 | — | — | — | — |
| 1918 | 202 | — | — | — | — |
| 1919 | 213 | — | — | — | — |
| 1920 | 211 | 155.8 | 101.0 | 222 | 226 |
| 1921 | 125 | 122.8 | 115.0 | 154 | 184 |
| 1922 | 132 | 112.6 | 107.8 | 131 | 168 |
| 1923 | 142 | 125.1 | 122.1 | 139 | 167 |
| 1924 | 143 | 117.3 | 105.3 | 141 | 168 |
| 1925 | 156 | 120.4 | 106.0 | 157 | 178 |
| 1926 | 145 | 119.6 | 98.0 | 147 | 192 |
| 1927 | 139 | 122.0 | 92.3 | 131 | 178 |
| 1928 | 149 | 124.4 | 102.9 | 141 | 186 |
| 1929 | 146 | 123.6 | 95.4 | 149 | 196 |
| 1930 | 126 | 107.8 | 86.1 | 117 | 162 |
| 1931 | 87 | 85.0 | 75.4 | 87 | 136 |
| 1932 | 65 | 65.3 | 61.1 | 74 | 102 |
| 1933 | 70 | 64.5 | 65.4 | 71 | 83 |
| 1934 | 90 | 75.6 | 66.7 | 86 | 98 |
| 1935 | 108 | 85.0 | 71.6 | 112 | 130 |
| 1936 | 114 | 94.5 | 75.3 | 107 | 136 |
| 1937 | 121 | 97.6 | 76.0 | 101 | 137 |
| 1938 | 95 | 85.8 | 65.5 | 99 | 132 |
| 1939 | 92 | 81.9 | 69.8 | 86 | 118 |
| 1940 | 98 | 88.9 | 74.0 | 89 | 115 |
| 1941 | 122 | 103.1 | 90.3 | 116 | 137 |
| 1942** | — | 115.7 | — | 147 | 161 |

*"Agricultural Outlook Charts, 1943," Bureau of Agricultural Economics, U.S. Department of Agriculture.
**Preliminary figures.

The statistics given above are portrayed in the following four graphs to give a more vivid realization of the periodic rise and fall of nation-wide price levels which control the prices which farmers may expect from their various products irrespective of the marketing systems involved:

*Prices Received by Farmers for Products as a Whole*
*1910–41*
(1910–14=100)

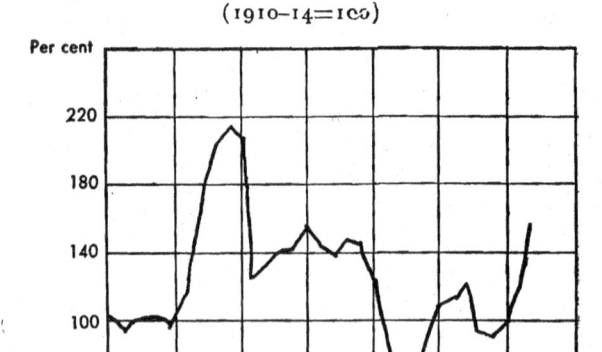

*Prices Received by Farmers for Dairy Products 1920–42*
(1922–41=100)

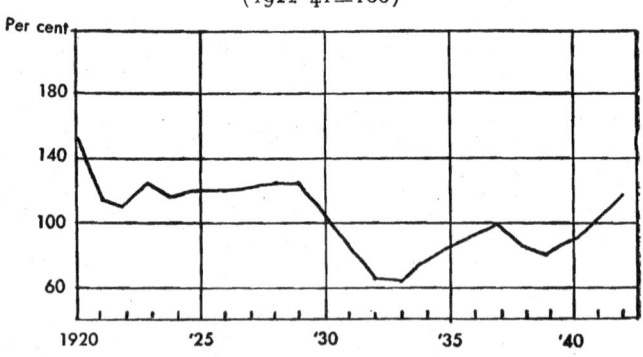

*Prices Received by Farmers for Chickens and Eggs 1920–42*
(1910–14=100)

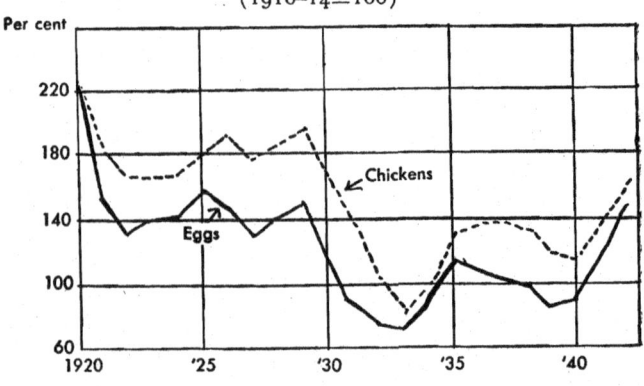

## MARKETING

*Prices Received by Farmers for 17 Vegetables for Fresh Market 1920–41*
(1924–29=100)

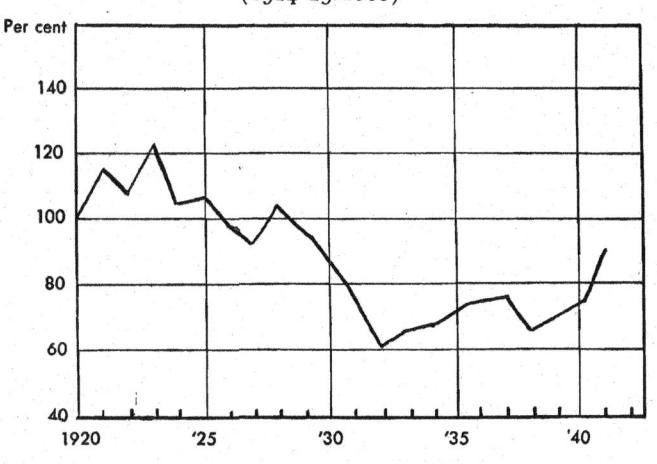

Since certain services and costs are inevitable, what are some of the factors in marketing that a farmer should consider and plan for in selling his products?

### SELLING DIRECT TO CONSUMERS

Obtaining a farm near a large town or city will make it possible to sell direct to the consumer. This will return to the farmer the highest net price for his products. Farmers who decide to go into the retail trade must expect to spend much of their time in preparation and delivery of the product. They must usually be good salesmen, study the demands of individual families, and know whether to extend credit.

Selling at retail from a farm entails considerable care in planning farm operations. Families who depend upon near-by farmers for goods will demand dependable service; there can be no excuse for poor quality or lack of products. Vegetable growers who have catered to consumers find that they must have a succession of crops in order to hold their trade; this necessitates making several plantings of the same products, such as sweet corn, lettuce, or beets. Regardless of weather, a farmer who has an egg route must be regular in his deliveries and his eggs must appeal to the customer.

On the other hand, in periods of depression a farmer selling at retail suffers less change in price levels than those selling in wholesale amounts.

### Quality of Products

On any market at a given time, there is usually a price range quoted for various qualities. In some cases, due to superior quality, a small percentage of goods will bring 10 to 20 per cent more than the going average price. Careful methods of growing may bring the additional price, or it may be that the way the product is packed or displayed contributes to the buyers' bids for the product. Inherent quality and local market demands are both factors that have an effect upon returns. McIntosh apples are favored in New York, with Staymens being desired in Philadelphia. Large eggs that have been candled and kept cool bring the best prices. On the Boston market long green asparagus brings a premium. Every farmer can afford to spend some time studying the needs of his near-by market and putting his products up in a way that meets those needs. Too often, farmers fail to consider the final consumer's reaction to their products, with a consequent loss in demand.

The table opposite illustrates two points: (1) the variation in price which may occur in one market on one day according to the variety of the product, and (2) the range in price for different offerings of one variety due to differences in quality as shown by daily price-groupings of "high," "low," and "mostly." In this case the quality range within each variety of beans had more effect upon the price than the differences in market demand for the different varieties of beans.

### Market Outlets

Although some farmers will be so situated that they can sell their products direct to the consumer, this will apply only to a few cases. The majority will, of necessity, use other outlets. Given the opportunity, most farmers would prefer to sell all their products at the farm for cash, and while this is done at times, it is not generally practiced. When a product is scarce, buyers may go direct to farms to obtain a supply to fill their needs, or some farmers may grow large enough quantities or produce goods of

## WHOLESALE PRICES OF SNAP BEANS—NEW YORK CITY MARKET—1942

|   | Bountiful |   |   | Plentiful |   |   | Valentine |   |   | Wax |   |   |
|---|---|---|---|---|---|---|---|---|---|---|---|---|
|   | High | Low | Mostly | High | Low | Mostly | High | Low | Mostly | High | Low | Mostly |
| July 1 | 1.50 | .50 | 1.15 | 1.50 | .60 | 1.00 | 1.75 | .60 | 1.25 | 2.00 | .25 | 1.35 |
| 2 | 1.50 | .50 | 1.25 | — | — | — | 1.50 | .75 | 1.10 | 1.75 | .50 | 1.50 |
| 3 | 2.25 | .75 | 1.75–2.00 | 2.25 | 2.00 | 2.00–2.25 | — | — | — | 2.50 | 1.00 | 2.00–2.25 |
| 6 | 2.00 | .50 | 1.25 | 1.50 | 1.00 | 1.25–1.50 | 2.00 | 1.00 | 1.25–1.75 | 2.00 | .50 | 1.50 |
| 7 | 2.00 | .25 | 1.00–1.50 | 1.25 | 1.00 | 1.00–1.25 | 1.25 | .60 | 1.25 | 1.75 | .50 | 1.50–2.00 |
| 8 | 2.00 | .25 | 1.25–1.50 | — | — | — | 2.25 | .75 | 1.25 | 2.00 | .50 | 1.75 |
| 9 | 2.00 | .85 | 1.00–1.50 | 2.00 | 1.50 | 1.50–2.00 | 2.00 | 1.00 | 1.25–1.75 | 2.50 | .50 | 1.75 |
| 10 | 2.50 | .50 | 1.25–1.75 | 2.25 | 1.50 | 1.50–2.00 | 2.25 | 1.25 | 1.50–2.00 | 2.50 | 1.00 | 1.50–2.00 |
| 13 | 2.25 | .50 | 1.25–1.75 | 2.00 | 1.25 | 1.25–2.00 | 2.25 | 1.75 | 1.75–2.25 | 2.00 | .50 | 1.50–2.00 |
| 14 | 2.00 | 1.00 | 1.25–1.75 | 1.50 | 1.00 | 1.00–1.50 | 2.25 | 1.25 | 1.50–2.00 | 2.50 | 1.00 | 1.25–1.75 |
| 15 | 1.50 | .60 | 1.00–1.50 | 1.75 | 1.25 | 1.25–1.75 | 2.00 | 1.13 | 1.25–1.75 | 2.50 | .50 | 1.50–1.75 |
| 16 | 1.50 | 1.00 | 1.25–1.50 | — | — | — | 2.00 | 1.13 | 1.50–1.75 | 2.75 | .75 | 1.25–1.75 |
| 17 | 2.00 | .50 | 1.25–1.75 | 1.50 | .75 | 1.25–1.50 | 2.00 | 1.25 | 1.50 | 2.25 | .25 | 1.50–2.00 |
| 20 | 2.00 | .75 | 1.25–1.50 | 1.38 | 1.13 | 1.13–1.38 | 2.00 | 1.00 | 1.25–1.50 | 2.50 | .75 | 1.00–1.75 |
| 21 | 1.75 | .50 | 1.00–1.25 | 1.25 | 1.00 | 1.00–1.25 | 2.00 | 1.00 | 1.00–1.25 | 2.00 | .75 | 1.50–2.00 |
| 22 | 1.75 | .50 | 1.00–1.50 | 1.25 | 1.00 | 1.00–1.25 | 1.75 | .75 | 1.25–1.75 | 2.00 | .75 | 1.00–1.50 |
| 23 | 1.50 | .50 | 1.00–1.25 | 1.50 | 1.00 | 1.00–1.25 | 1.75 | .50 | 1.25–1.75 | 2.25 | .50 | 1.25–1.75 |
| 24 | 1.50 | .40 | 1.00–1.25 | — | — | — | 1.75 | 1.00 | 1.25–1.75 | 2.00 | 1.00 | 1.50–2.00 |
| 27 | 1.75 | .25 | 1.25–1.50 | 1.75 | .75 | 1.25–1.75 | 1.75 | .75 | 1.25–1.75 | 2.50 | .25 | 1.25–1.75 |
| 28 | 2.25 | .75 | 1.25–1.50 | 1.75 | 1.50 | 1.50–1.75 | 1.75 | 1.25 | 1.25–1.75 | 2.00 | 1.50 | 1.50–2.00 |
| 29 | 2.00 | .50 | 1.25–1.50 | — | — | — | 2.00 | 1.00 | 1.50–2.00 | 2.00 | 1.00 | 1.75–2.00 |
| 30 | 1.75 | .50 | 1.00–1.50 | — | — | — | 2.00 | 1.00 | 1.50–1.75 | 2.00 | .50 | 1.50–2.00 |
| 31 | 1.75 | .25 | 1.25–1.75 | 2.00 | 1.00 | 1.50–2.00 | 1.75 | .50 | 1.50–1.75 | 2.00 | 1.00 | 1.50–2.00 |

sufficiently special qualities to make it desirable for buyers to go to their farms and make outright purchases. These cases will be the exceptions rather than the rule. Farmers, however, have certain choices in their market outlets, depending upon their location, the type of product they have to sell, and the market developments that have taken place in their community.

## Dairy Products

Dairy products need processing before they can be consumed, and, because of their perishable nature, receiving stations have been set up at country points in dairy regions to handle milk or cream. In the Northeast, where fluid milk is the main dairy product sold, there is no lack of country milk plants to which farmers can deliver or send their milk. A farmer may, if he is within close trucking distance to a town or city, establish a retail route and thereby include distribution as a part of his farming operations. This would add considerably to his income, but he would have to meet local health requirements applying to distributors and obtain a license before he could start deliveries. He must also bottle the milk, deliver it, and meet grade requirements of bacteria count and butterfat content.

In wholesaling his milk, the farmer may deal direct with private corporations or he may join a farmers' organization which will act as his bargaining agent in making contacts with dealers. In this latter case, his milk would be delivered to a near-by plant owned by the corporation that processes the milk, but actual sales terms would be made through the farmer's own association.

In the fluid-milk sections, the price the farmer receives for his milk is based on its actual use. In other words, the price quoted for Class 1 or fluid milk may not be the price that the farmer receives for all of his output. The price he actually has returned to him is known as a blended or composite price, made up from the actual amount sold as Class 1 or fluid milk, the amount the dealer sells as cream, and the amount sold for ice cream or other products. The farmer's final returns for the milk he takes to the plant will therefore vary according to the amount the dealer sells in different ways.

In the fluid-milk sections, health regulations play an impor-

## MARKETING

tant part in the production of milk on the farm and often delineate the areas which may ship milk to a certain market. Conforming to the health regulations as determined by the health officer of any city or state is absolutely essential on the part of any dairyman supplying milk to any market, and the health officer is supreme in enforcing health regulations.

More remote from cities, dairy products are delivered in the form of cream to plants making butter, and there is a large market for milk to go to condenseries and to cheese-making plants. Both private and cooperative concerns are established in the dairy sections of the United States usually available to any dairy farmer wishing to dispose of his product.

In the last decade there has been a definite trend toward public regulation in the marketing of fluid milk, with states passing milk-control acts and the Federal government setting up a plan whereby milk can be handled under a Federal market-agreement program. The question of a satisfactory price for fluid milk for farmers, on the one hand, and consumers, on the other, has been one on which there have been many investigations and much controversy. A large part of the cost of getting milk to the consumer is in the retail distributing end, and this is often as high as 25 or 30 per cent of the total cost of a quart of milk to the consumer. In some cities, particularly in the Central States, there have been experiments in giving discounts where several quarts of milk are delivered to one household, and there have been some trials in selling milk in gallon jugs in order to reduce delivery costs. In New York City, there has been a definite trend toward selling milk through stores in order to reduce delivery costs; approximately 60 per cent of all milk used there is sold through stores. Where labor costs are high and deliverymen are well organized, it will be difficult to reduce the cost of marketing milk.

### Poultry and Poultry Products

A generation or more ago, many farmers took their eggs to the country store and exchanged them for groceries. While this was an easy way for the farmer to get rid of his eggs, it was not satisfactory either to the consumer or the farmer, partly because the eggs were not given enough attention at the farm and were not delivered regularly, and partly because the store man was not equipped to handle them efficiently. He had no well-equipped place to store them and seldom graded them. In some areas the farmer would sell his eggs to a country buyer who came around weekly to collect a supply, or the farmer would ship them to a

*Wire basket, used for collecting eggs, permits free circulation of air and thus facilitates rapid cooling.*

large town or city to a commission merchant. Those farmers who were near to cities often established egg routes or delivered eggs direct to retail stores.

As the demand for improved quality of eggs increased, more attention was paid to the care of eggs upon the farm, and facilities for handling eggs were established in country locations. It was therefore a natural development that plants for receiving, grading, shipping, and selling eggs have been established in many sections of the country. Producers became interested also, and a farmers' cooperative known as Pacific Egg Producers built plants

## MARKETING

for receiving eggs on the Pacific Coast and set up their own sales organization to sell eggs on eastern markets. This cooperative made a specialty of grading and standardizing the eggs they received from their own farm members. The eggs were assembled in warehouses on the West Coast, then candled, graded, and packed according to buyers' demands, and given rapid transportation to the big cities of the East. Because of the high standard of quality and the ease of handling a graded product, dealers in

*Poultrymen's Cooperative Egg Auction Market*

eastern markets turned to these eggs and distributed them to many retail stores in this big consuming section. Eggs from the West Coast then began to be quoted on the New York Market at a premium over near-by eggs. Northeastern poultry producers noticed this situation and established cooperative egg and poultry organizations of their own in poultry sections and within easy trucking distance of large cities.

These auctions or country egg stations receive eggs from farmers in case lots, sort them for grade and weight, and then auction them off to buyers. The charge for these services is then deducted from the selling price in the returns made to the farmer. Many of these plants go further in the marketing operation, and will also deliver eggs to retail stores—sometimes even candling and packing them in cartons of a dozen, ready for retail distributors.

## PRODUCING FOR PROFIT

In one of the eastern states, approximately 50 per cent of all eggs produced are now marketed through cooperative organizations. In some of the states more distant from large consuming centers, even a greater proportion of the eggs produced is handled through cooperatives.

Regardless of whether these services in connection with handling eggs have been performed by cooperatives or private concerns, there has been considerable improvement in the care and grading of eggs in recent years, compared with the methods used formerly.

Processing methods, shifts in population, price fluctuations, and transportation conditions have made considerable difference in the receipts of eggs on various markets. Much of the supply of eggs from the Pacific States used to be shipped to eastern markets but is now being consumed in cities in California, Oregon, and Washington, and large quantities of eggs in the Central West are now being dried.

Regardless of such developments, eggs of high quality have been in such demand that retail routes have continued, and many farmers, if they are close to cities, still choose to deliver eggs to consumers, or, if opportunity offers, they deliver regularly to retail stores.

Outlets open to farmers selling eggs will include (1) consignments to large city receivers or commission men, (2) delivery to near-by country receiving and shipping plants, either cooperative or private, (3) selling to country collectors who come to the farm and pick up the eggs, (4) direct delivery to retail stores, (5) direct delivery to consumers. Each farmer will want to study his own conditions, estimating the returns he would receive from the various types of available outlets, before deciding how he should sell.

One word of caution should be offered concerning the consignment of eggs to commission merchants in distant markets. A number of the farm papers run service bureaus and print letters from their readers, of which the following is typical: "I have a check for eggs given me by ———— Company which was returned from the bank marked 'Insufficient Funds.' Can you help me collect it?" The service bureau page of many of these farm papers tells readers that the firm they inquire about is no longer in existence or that it is not a reliable concern, and

## MARKETING

cautions farmers to look up the reliability of commission merchants before making shipments to them. In a number of states, firms engaged in buying and selling agricultural products have

*One method of maintaining egg quality is represented in this egg room, constructed as a result of banker-farmer cooperative effort. At left is cooling cabinet, taken apart, to show construction. Operations are based on the principle of forcing cool, moist air into cabinet and around freshly gathered eggs. Note section of ordinary eaves trough (top), equipped with a float valve to maintain a constant water level, so that moisture may be absorbed by feed sacks hanging from trough. These sacks serve as wicks and, when heavy with moisture, serve to raise humidity in the egg room. Cost of this homemade installation is about $45.00.*

to be licensed and bonded before starting in business, and farmers should check with their State Departments of Agriculture if they desire to find out whether the firms to which they are considering sending their products have complied with the State law or have been issued a license, before dealing with them.

## PRODUCING FOR PROFIT

Poultry is not sold regularly and continuously, as are eggs, but quite often the same outlets are available to farmers for the sale of their broilers, roasters, or fowl. Many of the plants set up to handle eggs are also equipped for poultry. The farmers also consign their supplies to receivers of live products or sell them to buyers who come to their farms and pick them up. New York City is one of the largest live poultry markets in the country and, because of its unorganized condition, the city department of markets has established a central live poultry terminal in Brooklyn. Both truck and rail receipts are handled at this new terminal, where they are unloaded and sold under certain rules and regulations. It is hoped that by concentrating the arrivals and unloadings at one point, with some supervision by government inspectors, better market conditions in the reporting of prices can be made and costs of selling and handling of the poultry can be reduced.

### Fruits and Vegetables

Markets for fruits and vegetables will vary all the way from roadside stands, through city farmers' markets, to the big receiving terminals in such cities as New York, Boston, Philadelphia, Detroit, Chicago, or Los Angeles. Previous to the first World War, much of the fruit and vegetable supply of consumers was raised locally, and a considerable quantity was sold on local farmers' markets. With the development of the refrigerator cars, much new land was opened in the South and West, and shipments of out-of-season vegetables to northern markets increased many fold. With the rise of the volume of shipments by rail, facilities for the special handling of receipts by rail were built. Such terminals were established in a number of cities, including those mentioned above, almost entirely for fruits and vegetables coming to market by rail.

In the last two decades, other developments have taken place, such as delivery of fruits and vegetables by truck and the growth of chain stores as distributing units in the fruit and vegetable business. These changes have been significant and have caused a good deal of confusion in the handling of the fruits and vegetables. In the city farmers' markets, which formerly attracted a large number of farmers and a large number of independent retail store buyers, the change has been quite apparent, and there

## MARKETING

are fewer buyers there than formerly. Some farmers within trucking distance of cities sell and deliver direct to chain-store warehouses. Thus, it has developed that rail receipts come to the railroad terminal, trucked-in supplies come to warehouses at a different point, and in some cases the farmers' market is in still a different location. This has caused difficulties because buyers have had to go to two or three markets to purchase their goods,

*Wholesale produce markets in such large cities as New York, Philadelphia, and Boston are antiquated, congested, and inadequate; in such markets costs of distribution are increased unduly to the detriment of producers, consumers, and middlemen.*

and it has been difficult to keep costs of handling fruits and vegetables low, because of the necessity of transporting the products from one part of the city to another to assemble them. An attempt has been made both in the country and the city to improve this situation. At country points, assembly markets have been organized in the southeastern states and also in the northeast section; and in the city markets, recommendations have been made to consolidate the sale of products that arrive by rail and truck into one terminal market.[3]

Developments in processing may have a considerable effect in the future on the marketing of fruits and vegetables. There has been a considerable growth in the sale of quick-frozen products, such as berries and certain types of vegetables, where the process-

[3] "The New York Fruit and Vegetable Markets," Bureau of Agricultural Economics, U.S. Department of Agriculture.

ing is done near the point of production, thus obviating the necessity for the vegetables to be handled through the usual city terminal markets. All fruits and vegetables cannot be successfully frozen, but with increased developments in freezing and dehydrating of fruits and vegetables, different methods of mer-

*Traffic jams in large city wholesale markets not laid out or equipped to receive truck shipments efficiently waste time, increase distribution costs, and contribute materially to impairment of the quality of fresh fruits and vegetables.*

chandising are bound to occur in the marketing of these products.

Outlets that are open to fruit and vegetable growers will include (1) delivery to country buyers or country receiving and shipping plants, (2) consignment to receivers or commission merchants in a near-by or distant city, (3) sale or delivery to large chain stores, (4) sale by farmer on a city farmers' market, (5) sale to retail stores, (6) sale through roadside stands, (7) delivery direct to consumers.

Shipment by a farmer to a commission merchant or receiver should be made only after the credit or the reputation of the dealer has been ascertained. As is the case with other products, many states have passed laws requiring that commission merchants handling fruits and vegetables must be licensed or bonded before they can do business. Reputable buyers, however, have

## MARKETING

been in business in many city markets for long periods and are able to give satisfactory returns to their producers. The commission merchant charges a certain commission on the sale of the products—usually 8 or 10 per cent; or, if the products are sent to a receiver or outright buyer, the receiver pays the farmer

*Washing celery and radishes on a truck farm prior to shipment to market.*

an f.o.b. price at the country point or a price delivered at the city. The difference in the two sales methods is one of ownership of the goods. Under the commission method, the commission merchant acts as the agent of the grower, charging a fee for his services. The receiver, on the other hand, makes an outright purchase of the goods, either at country points or at a delivery point in the city.

### GOVERNMENT AIDS IN MARKETING

Because of the perishable nature of the products, the distance separating farmers from city markets, and the fact that many types of agencies, both buyers and sellers, are concerned in the business of marketing farm products, there has been considerable demand that the government set up certain services in connection with the marketing of agricultural products. Farmers, in selling their crops or livestock, and buyers who wish to keep in touch with conditions are able to get much statistical information on the size of the crop, the movement of crops or livestock to market, the market prices prevailing in different places at any particular time, and reports showing the grades of the product being sold.

## Crop and Market Reports

The Federal government issues information on the size of the important crops and livestock being grown in the United States, estimating the probable amount to be harvested or produced, and in December giving an estimate of actual production. During the growing season, not only can farmers find out the estimate of the crop to be harvested, but, when movement to market starts, they can obtain regular reports on shipments from country points, arrivals at various markets, and the prevailing prices on those markets. Farmers can therefore follow a commodity like potatoes, knowing from day to day the number of cars loaded in Maine or in Kansas or in Virginia, the number of car loads being received at New York, Chicago, or Detroit from those shipping areas, and the prevailing price per hundred pounds at these markets at whatever time the sales are made.

In all of the larger cities of the United States, either Federal market reports or joint reports by Federal and state officials are issued. These cover prices on livestock products, poultry and dairy products, and fruits and vegetables. Special market reports are issued on grains, naval stores, cotton, and some other products.

Information concerning any of these reports may be obtained by writing to the Agricultural Marketing Service of the U. S. Department of Agriculture. Careful study of these reports will repay those having products to sell by making possible the use of good judgment as to when, where, and how to dispose of their products to the best advantage.

In order to keep everyone, buyers and sellers alike, equally well informed about supply, demand, and prices of goods on the various markets, farm market radio reports are increasing in importance and scope. Market reports come out as early as 6:30 in the morning on the eastern markets and are given during the day on livestock and grain markets, in order that quick dissemi-

## MARKETING

nation of such information may be made throughout the country. This helps prevent scarcities in one part of the country and gluts in others.

### Inspection Services

In order that business can be done more efficiently and with less time between two parties at a distance, the Federal government has set up official grades or standards for farm products, so that a buyer and a seller widely separated from one another can make a contract on the basis of a description of the product. The Federal government will, upon request of a buyer or seller and for a nominal fee, inspect a carload of potatoes either at shipping or receiving points on the basis of the United States grade, and a certificate of inspection will be issued against the carload or lot examined and a sale can be made by description, using this grade as a basis of sale without the buyer seeing the goods. Buyers and sellers now make contracts for exchange of goods by letter or by wire, based on United States grades and accompanied by an inspection certificate signed by a Federal inspector. The establishment of this service has reduced some of the costs of marketing and is also used as a basis for the settlement of disputes. The establishment of grades for farm products ranges through most of the types of articles grown on farms and includes strawberries, asparagus, apples, butter, wheat, cotton, eggs, hay, honey, maple syrup, and a large list of products.

### Improving Marketing Efficiency

Many public efforts are concentrated on trying to reduce the spread or the margin between the price paid to the farmers and the price paid by consumers. The Federal government and the various Land Grant Colleges conduct considerable research into the field as the necessary first step toward building any program. Government services have been developed for issuing market reports, information on stocks available, and the volume in cold storages. In addition, both state and Federal governments regulate to a considerable extent the conditions of sale of many of our perishable products. All commission merchants handling fruits and vegetables must be Federally licensed, and in many states they must be licensed and bonded. Commodity markets,

## PRODUCING FOR PROFIT

such as the Chicago grain exchange, also operate under the conditions of Federal control. It is likely that more rather than less government effort will be directed toward marketing practices.

Private concerns, as well as cooperatives, are constantly seeking better ways to market food products. Reduction in assem-

*Production Point Farmers' Cooperative Auction Market.*

bling costs, the hauling of larger loads, the reduction of waste in spoilage, reduction in the amount of labor in getting products from one point to another, are factors that both of these groups are constantly studying.

### Production and Marketing

Experienced farmers have long recognized that, as individuals, they necessarily must accept what the markets will pay for their products. There are exceptions, to be sure, such as the growers who sell from roadside markets and the farmers who have successfully developed profitable outlets for products mailed or otherwise shipped direct to consumers in town. But among some six million American farmers, they are relatively few in number. The farmer, in general, still takes what the market offers. Sometimes he likes it, sometimes he doesn't.

This is a fact that the beginner in farming might well ponder. This applies with special emphasis to the prospective farmer who has thought out a scheme, let us say, for marketing cellophane-wrapped eggs in fancy packages. Or perhaps it is some other supposedly new formula, expounded by a city man inexperienced in agriculture, for hitting the jackpot on the farm. On occasions, the beginner has worked out a unique and profitable plan of marketing, but more often too much emphasis has been placed on merchandising before learning the fundamentals of efficient production.

Here is as good a point as any to call attention to the fact that efficient marketing must begin with efficient production. The fruit grower who is unable to control insects and diseases will, at harvest time, have a disproportionately large amount of wormy and otherwise unattractive fruit to take to market. No amount of fancy packaging will overcome the handicap of low-quality fruit, as will be convincingly demonstrated by the price it brings on the market. The same lesson holds with respect to other products, whether they be eggs, milk, honey, cream, butter, or vegetables. The man "who can sell," if he is to succeed on the farm, must also be able to produce.

Always remember that special grading and fancy packaging undertaken in connection with merchandising programs designed to obtain a large share of the consumer's dollar call for the expenditure of considerable time and money. This represents time taken from production efforts, money diverted from possible enlargement or strengthening of the farm as a productive unit. Exactly how to achieve a desirable balance between time and money spent in production and marketing is obviously a separate problem on every farm. But it is a problem deserving of serious attention, especially in the case of the beginner who is certain that selling direct to the consumer is a sure formula for success.

If the prospective farmer understands at the outset that he probably will have to take what the market offers, he will at least know of a condition that confronts farmers generally. Then if he will make a close study of the usual farm marketing practices followed in the area where he intends to locate, he will at least have an understanding of the methods that most of his prospective neighbors have found best suited to their needs.

Such a study may reveal the existence of a thriving farmers' cooperative selling organization, in which case the advantages of joint selling efforts are available. Farmers in all sections of the United States have vastly strengthened their bargaining position in the nation's markets through cooperatives, and membership in, and active support of, such organizations is recommended.

## PRODUCING FOR PROFIT

### Suggested Readings

Economics of Cooperative Marketing, by Henry H. Bakken and Marvin A. Schaars. (McGraw-Hill Book Company, New York.)
The Economics of Marketing, by Hugh B. Killough. (Harper and Brothers, New York.)
High Quality Eggs for Illinois Markets, by H. H. Alp. (*Circular 494,* University of Illinois, Urbana, Ill.)
Marketing Agricultural Products, by F. E. Clark and L. D. H. Weld. (The Macmillan Company, New York.)
The Marketing and Distribution of Fruits and Vegetables, by Motor Truck. (*Technical Bulletin 272,* U. S. Department of Agriculture, Washington, D. C.)
Merchandising Fruits and Vegetables, by W. A. Sherman. (Shaw Publishing Company, Washington, D. C.)
Preparing Poultry for Market. (*Extension Leaflet 11,* Massachusetts State College, Amherst, Mass.)
Quality Eggs. (*Circular 160,* Massachusetts State College, Amherst, Mass.)
A Study of Fluid Milk Prices, by John M. Cassels. (Harvard University Press, Cambridge, Mass.)
Who Gets Your Food Dollar? by Hector Lazo and M. H. Bletz. (Harper and Brothers, New York.)
Wholesale Markets for Fruits and Vegetables in 40 Cities. (*Circular 463,* U. S. Department of Agriculture, Washington, D. C.)

# ROADSIDE MARKETING

## W. R. Stone

Not all farms are adaptable to the successful operation of a roadside market. The experience of those who have operated roadside markets and those who have studied methods of direct sales from producer to consumer indicates that three fundamentals should be considered before launching into this specialized method of marketing: (1) the farm must be located near a large residential area; (2) soil conditions must be adaptable to a wide range of crops; and (3) one or more members of the family should have a sound knowledge of production methods, ability to organize, possess sales ability and an abundant supply of patience.

Just good, plain common sense indicates that a farm located

## MARKETING

close to one or more residential towns has greater possibilities of attracting trade than one located 15 or 20 miles away from a thickly populated area.

Many farmers who pioneered in roadside marketing based their venture on attracting transient trade. They soon found that little dependence could be placed on customers who just happen to stop and buy.

The most successful roadside market operators are those favorably located on the edge of a residential area. The actual

*Proximity to well-traveled road and ample parking space are essential to the success of a farm roadside market.*

site or location of the stand is of great importance. It should be near the highway. Most states have highway laws which prevent locating the stand so close to the highway that it is likely to interfere with the orderly movement of traffic. It is well to check your highway law.

Ample parking facilities should be provided, preferably alongside of the market stand or at the rear. This area should, of course, be leveled and a heavy coating of cinders or similar material placed over the entire parking area. Some method of marking to provide efficient use of space and ease of parking is also helpful.

There are a great many different types of roadside buildings. Length of season the market will be open and estimated volume of sales will largely determine building type. A simple framework entirely or partially enclosed on three sides and tightly roofed, with the open side facing the highway, has proved satisfactory for sales during the summer and early fall months. If the market is operated throughout the year, more substantial construction is required. Full-length movable sash in the front has been used advantageously on this type of building. It should be insulated and provision made for heating.

Regardless of the type of building constructed, the tables, bins, and shelves should be set up to secure maximum display of sales

*Effective Display at a Farm Roadside Market*

and efficiency in handling. Either attached to the stand or located near-by should be a room or building where surplus products can be kept in a fresh condition. Usually, this is the place where vegetables are graded, washed, and made ready for the trade. In some states, the extension agricultural engineer of the State Agricultural College has plans for different types of construction.

It is important to check on zoning ordinances and to consult with the building inspector for the borough or township in which the farm is located before actual construction is started. The building should be painted annually and kept neat and attractive at all times. It is remarkable how a few ornamental plants properly placed will still further improve appearance.

A sound knowledge of production methods is basic for the successful operation of the roadside market. A great many factors enter into this phase of the enterprise. Experience again indicates that greatest success is obtained where a wide variety of crops is grown and placed on sale throughout the season.

## MARKETING

Some soils are not adaptable to growing a wide range of crops. Cold, heavy, clayey soils are particularly difficult to handle for growing crops. A sandy loam soil which warms up rather quickly and can be worked even during wet weather is considered ideal.

Planning is a most important phase of this enterprise. The operator must know those crops which can be grown and those which will not thrive under his soil conditions. He must know the products the trade wants and when they are in greatest demand. It is no simple task to outline and carry out a plan which will provide the market with an ample supply of vegetables required throughout the major portion of the season. The poultryman must do some careful planning to have plenty of eggs or meat when wanted. Several years of trial and error may elapse before a really satisfactory plan is hit upon.

The vegetable producer should also keep in mind growing those vegetables of highest quality and make ample provision for liming, fertilizing, insect and disease control, and labor requirements of the crops to be grown. Your County Agricultural Agent can be of assistance relative to crop rotation, companion cropping, succession crops, and other cultural practices.

Some roadside marketers believe it helps to have crops growing or poultry houses located so that the customers can see them from the highway or from the stand. Such conditions indicate to the customer that this is a bona fide farm, where the vegetables, eggs, or other commodities are actually produced.

Every effort should be directed toward supplying the consumer with farm products of highest quality and in the freshest possible condition. Some sweet corn growers pick corn at frequent intervals during the day so that the customer can secure it in as fresh condition as if it were picked from his own garden. Harvesting vegetable crops early in the morning and keeping them in a cool place, gathering eggs frequently and keeping them cool, are just two of many practices which help keep quality and freshness at a high level.

The roadside market operator who grows products of high quality, attractively displayed and courteously sold to the consumer, continues to conduct his business on a profitable basis in spite of changing economic conditions. A superior product to that which reaches the consumer through the usual channels will draw trade. There is no reason why the customer should burn his own gasoline and use his own tires to come to the farm for just ordinary foodstuffs.

In many states there are regulations pertaining to public health which apply to a retailer, but which the farmer has not had to consider where his products are sent to market. It would be well, therefore, to check with your local board of health concerning any regulations which might affect the operation of your roadside stand.

The county sealer of weights and measures should also be consulted to have him specify the type of scales or measures which are legal. He can explain laws and regulations which may save you trouble later. Package types should conform to these requirements.

This matter of packages plays an important part; generally speaking, they should be new, clean, serviceable, permit maximum display of the products they contain, and be suitable for the customer to take home in his car with a minimum amount of upsetting.

Standardization of packages and grading, tied in with standardized prices, is considered a good business policy; for example, if the customer knows that he can get a good grade of his favorite variety of apples in a 16-quart basket for $1.00 throughout its market season, he knows how to adjust his marketing and his finances better than if the price fluctuates frequently and the package type is changed from time to time.

One very successful grower maintains standard prices during periods of scarcity, even though he could sell on the wholesale market at a price higher than the standard price at his roadside market. He claims he more than recovers his losses by maintaining standard prices when commodities are plentiful. Some other growers argue that it pays best to vary prices in line with the abundance or scarcity of the product.

Most operators believe that they are entitled to a price at least as high as that charged by greengrocers and others selling farm products to the retail trade. Occasionally, when there is a large supply on hand, they may reduce prices on those products for a short period as a "special." The customer who buys principally on price often will be attracted by second-grade products. Price is not an easy thing to determine. Cost of production records,

## MARKETING

wholesale market quotations, and the reaction of the trade help the stand operator to determine his prices.

It has been found that town and city folks like to combine a business and pleasure trip when they get in their cars to go to a farm where a friendly rural atmosphere prevails. Recognition by the salesman at the stand of steady customers helps a lot. This does not necessarily mean that he must know and remember the name of the customer, but it helps if he recalls some incident or previous conversation with that individual.

Extra count—an apple or a peach handed to a youngster, a free glass of cider at the farm—has proved valuable in holding the trade by creating a friendly atmosphere.

There is a wide variation of opinion on the question of advertising. Some feel that they should have a standard ad regularly in one or more local papers in those towns where the trade resides; others feel that an occasional ad indicating that sweet corn, peaches, or some product which serves as a drawing card to the trade is now ready constitutes sufficient advertising.

Many roadside marketers think and talk about preparing a list of their customers so that they can send a postal card when certain products are ready. It is hard to find anyone who has carried out this plan. It usually happens that some time before the crop is matured some customers will call the stand operator to learn when the crop will be available.

Printed recipes placed in bags of produce or handed to customers in all cases have resulted in increased sales. The farmer can get the recipes from home demonstration agents and have them printed.

A source of never-ending discussion among roadside marketers relates to limiting sales to own-grown products versus supplementing these products with some purchased from the market, the largest number agreeing that it is necessary to purchase some crops, particularly early in the season, to supplement the sale of their own-grown products; for example, the grower who has a fine lot of lettuce to offer to the trade finds that it helps move that lettuce if he has some tomatoes, cucumbers, or other crops frequently used in salads, even though they are known to be out of season in that particular area at the time. Their opinion is that most customers would prefer to buy all of their vegetables at the one stand, rather than to purchase just those things grown

## PRODUCING FOR PROFIT

on the home farm and go somewhere else for products which are out of season.

There are still some roadside market folks who limit their sales to products grown on the home farm. Seasonal variations may make it necessary for even these folks to secure some high-quality product from a neighboring farm when their own product is insufficient to supply trade needs.

Generally speaking, the heaviest marketing takes place from 4:00 to 7:00 P.M. on week days. Saturday and Sunday business is many times greater than during the week.

In the spring or early summer, green vegetables, peas, and strawberries draw the trade; later, sweet corn, tomatoes, peaches, apples, and cider. In the metropolitan area, near New York City, heaviest roadside market activities occur between the middle of August and Thanksgiving.

In this outline an attempt has been made to emphasize the importance of consideration by the farmer or prospective farmer of certain factors which experience has shown play an important part in the successful operation of a roadside market.

Location close to heavily populated areas, attractive and convenient settings for the market building, a knowledge of production methods and the use of good, sound business sense in the actual marketing of the products have been stressed. No doubt other important factors may enter into the question of determining whether this type of marketing could be successfully followed in your case. It is a business which requires careful management, long hours, and plenty of patience. A successfully conducted roadside market provides the operator with the highest possible remuneration for his farm products.

### Suggested Readings

Plans for a Roadside Market, by M. C. Bond. (New York State College of Agriculture, Ithaca, N. Y.)

Roadside Marketing of Agricultural Products by Ohio Farmers, by C. W. Hauck and H. M. Herschler. (Ohio Agricultural Experiment Station, Wooster, Ohio.)

Roadside Markets. (*Leaflet 68-L*, U. S. Department of Agriculture, Washington, D. C.)

Roadside Market Survey, by H. S. Kahle. (New York State College of Agriculture, Ithaca, N. Y.)

Printed in Great Britain
by Amazon